I0036976

Early Praise for *Write Better with Vale*

I'd never encountered Vale before reading this book, but I'm much better informed now. Hogan's writing, as always, is clear, informative, and most importantly, actionable. I now plan to implement Vale both at my organization and for my personal projects.

➤ **Steve Hill**
Senior Developer, Resource Guru

I would recommend this book to technical writers or anyone in a documentation-heavy role who is looking to enhance their existing workflow, enforce consistent style, and improve the overall quality of their written content. It's a practical and effective guide.

➤ **Abhigna Nagaraja**
Staff Software Engineer

Write Better with Vale

Automate Your Style Guides and Lint Prose Like You Lint Code

Brian P. Hogan

The Pragmatic Bookshelf

Dallas, Texas

Pragmatic Bookshelf

See our complete catalog of hands-on, practical,
and Pragmatic content for software developers:
https://pragprog.com

Sales, volume licensing, and support:
support@pragprog.com

Derivative works, AI training and testing,
international translations, and other rights:
rights@pragprog.com

The team that produced this book includes:

Publisher:	Dave Thomas
COO:	Janet Furlow
Development Editor:	Susannah Davidson
Copy Editor:	Corina Lebegioara

Copyright © 2025 The Pragmatic Programmers, LLC.

All rights reserved. No part of this publication may be reproduced by any means, nor may any derivative works be made from this publication, nor may this content be used to train or test an artificial intelligence system, without the prior consent of the publisher.

When we are aware that a term used in this book is claimed as a trademark, the designation is printed with an initial capital letter or in all capitals.

The Pragmatic Starter Kit, The Pragmatic Programmer, Pragmatic Programming, Pragmatic Bookshelf, PragProg, and the linking *g* device are trademarks of The Pragmatic Programmers, LLC.

Every precaution was taken in the preparation of this book. However, the publisher assumes no responsibility for errors or omissions or for damages that may result from the use of information (including program listings) contained herein.

ISBN-13: 979-8-88865-181-0
Book version: P1.0—December, 2025

Contents

Acknowledgments

This is my 14th book. When I decide to write about something, it's because I am deeply excited about the topic. I get paid to write, whether it's public-facing documentation or internal plans and proposals, so if I'm going to take on additional writing, it has to be something I care about. Vale is one of those tools I can't live without. It's a crucial component of my workflow whether I'm managing product documentation or writing blog posts or articles. I'm deeply grateful to Joseph Kato for creating this tool and being so responsive to user feedback over the years. Thank you, Joseph, for dedicating so much time and effort to creating a tool that makes us all better at our jobs.

Thank you, Dave Thomas, for continuing to give me, and authors like me, an outlet to share our knowledge and experience with the world. We're all better off because of the environment you and Andy Hunt created for us to publish.

Thank you, Susannah Davidson, for editing this edition. We've worked on many books together, and I learn something new each time. We make a great team.

Thank you, Steve Hill, Abhigna Nagaraja, Stefan Turalski, Zulfikar Dharmawan, Matthew White, and Raymond Machira, for your early technical feedback on the book. You found some serious show-stoppers and places where I needed to make things clearer. This book is much better as a result. And thank you, Margaret Eldridge, for your unique perspectives on the book that pushed me to rethink some explanations and approaches I took. I added several new small but important paragraphs based on your review.

Thank you, Matt Bernier, Mason Egger, Tim Falls, Alex Garnett, Jordan Glasner, Dee Klieb, Sam Linville, Amelia Mango, Brian MacDonald, Walter Poupore, Dave Rankin, Sydney Rossman-Reich, Lisa Tagliaferri, Erich Tesky, Candace van Oostrum, Larah Vasquez, Cully Wakelin, and Chris Warren, for your support.

Thank you, Ana and Lisa, for your love and inspiration.

Finally, thank you, Carissa, for your love and encouragement through this project and all the others. You help me every day to be the best I can be, and I couldn't ask for a better partner.

Preface

Content is a crucial part of your product's user experience, whether it's your product documentation, blog, use cases, or even customer education material. Your technical content is often the first place technical people go to evaluate a product, and clear, clean, and consistent content creates a great experience.

Style guides lay out the rules for that clarity and consistency. Whether you're writing for your English class or your company's blog, it's a good bet there's a style you're supposed to follow, like making sure all your headings use title-casing or that you avoid certain words. Sometimes the style guide lives on a series of web pages or in an internal knowledge base. Other times, it's held in the headspace of your senior editors.

If you write code, you've no doubt come across "code linters," tools that enforce code style. You can get the same experience for your prose.

Vale[1] lets you turn a style guide into a set of rules that ensure your writing is consistent. Vale is a prose linter. You can use Vale to check for spelling errors, typos, heading consistency, profanity, grammar issues, and much more. You can hook Vale into your workflow and have it lint your writing while you work and incorporate it into your continuous integration process. Using Vale, you can automate your style guide.

Here's how it works. You define rules that tell Vale what to look for. You can tell Vale to flag certain words or phrases, suggest substitutions to ensure authors use words consistently, or even ensure your content meets a certain word count. Vale parses your content into components like headings, lists, paragraphs, and sentences, and applies your rules. That means you can look for things in the entire document, or focus on specific parts, like the number of characters in your front matter's description field, or whether your headings use title case. It knows to ignore things like code blocks or URLs, so you don't have to worry about false positives in technical documentation.

---

1.   https://vale.sh

Best of all, you don't have to start from scratch. If you based your style guide on Microsoft or Google's style guides, you can import Vale's implementations of those guides, or many others, into your project and extend them or change them to suit your needs.

In this short book you'll get hands-on with Vale, customizing its features, using rules from the community, and building your own style you can use to improve the user experience for your documentation.

What's in This Book?

This book is your guide to setting up and using Vale on your content project. You'll get comfortable with Vale's features as you implement a style guide against a collection of inconsistent Markdown documents. You can use Vale with other formats like reStructuredText and AsciiDoc, but in this book, you'll focus on constructing style rules and using Vale in a practical use case.

If you use a different format for your own content, you'll find that most of the examples in this book will still work for you, although you'll have to modify some regular expressions when working with rules that look at markup rather than prose. Those cases are rare. Vale's documentation will help you work with other formats.

In Chapter 1, Getting Started with Vale, on page 1, you'll start out by installing Vale and using it to look for basic spelling issues in Markdown and HTML documents. You'll explore some of Vale's command-line options and build a minimal configuration.

In Chapter 2, Using Existing Styles, on page 15, you'll incorporate some existing community style rules into your configuration as you work to improve a small collection of tutorials. You'll make two overlapping style guide implementations work together, and then you'll make content more inclusive and readable, all with preexisting rules.

In Chapter 3, Building Your Own Style, on page 27, you'll build your own rules from scratch, along with your own vocabulary, so you can make sure that your company's technical jargon not only is not flagged as spelling errors but also is represented correctly and consistently.

Finally, in Chapter 4, Integrating Vale into Your Workflow, on page 53, you'll integrate Vale with existing tools, including your editor, your GitHub workflow, and formats that Vale doesn't directly understand. And you'll create a package for your style so you can share it with other projects or other teams.

This book isn't meant to replace Vale's official documentation, so you won't find every feature covered. This is a focused guide that helps you implement Vale in your project. When you're done, you'll be able to confidently investigate other Vale use cases.

Who Should Read This Book

This book is for anyone who works in text-based documents who wants to automatically check that content against a style guide. Anyone with experience with Markdown and command-line tools will be able to follow the examples to build a style guide. If you can write text in a text editor and run some basic commands in the terminal, you'll have no problems.

Technical writers who use "docs-as-code" workflows will feel at home, as will software developers who use linting tools like ESLint.[2]

What You Need

You'll need your trusty text editor, some command-line knowledge, and some idea of the style guide you'd like to implement.

You should also have some experience with regular expressions. You'll see and write lots of regular expressions in this book. You'll find Regex101.com[3] helpful as you write and debug your rules.

Where to Go for Help

Visit the book's web page[4] and the book's companion website[5] for example code or to report an issue with the book's examples. You'll also find a link to the book's discussion forum, where you can interact with other readers.

Vale is a powerful tool that helps you write better quality, consistent documentation. Ready to get started?

---

2.   https://eslint.org/
3.   https://regex101.com
4.   https://pragprog.com/titles/bhvale
5.   https://thevalebook.com

CHAPTER 1

Getting Started with Vale

Whether you're working with content in HTML, Markdown, AsciiDoc, or another format Vale supports, configuring Vale to lint your project follows the same initial steps: you need to install the Vale program and create a configuration file that outlines what rules to apply and what types of files to apply them to. You can get Vale running and checking your content for spelling errors and repeated words with almost no configuration thanks to the built-in rules and dictionary Vale provides.

To get comfortable with the basics, you'll install Vale's command-line tool, create a Markdown file with some deliberate mistakes, and then configure Vale to find those mistakes so you can fix them. Then you'll work with an HTML file to understand how Vale handles different file types and dive into some configuration options that determine which rules apply.

First, you'll need Vale running on your machine.

Installing Vale

Vale works as a command-line tool. You can install it through package managers, build it from source, or download a release binary. You can also use a Docker image.

On macOS, install Vale with Homebrew:[1]

```
$ brew install vale
```

On Windows, install with the Chocolatey[2] package manager:

```
$ choco install vale
```

---

1. https://brew.sh
2. https://chocolatey.org/

On Linux, you can install the Snap[3] package:

```
$ snap install vale
```

These methods ensure you get the latest version and that Vale is available on your system path so you can use it everywhere on your system.

You can also install one of the precompiled releases[4] from GitHub. Download the release, unzip the files, and place the vale executable in a directory on your PATH so you can access it everywhere.

Once you've installed Vale, open a new terminal window and execute the following command to run Vale and return the version number:

```
$ vale --version
vale version 3.13.0
```

You'll continue to use Vale on the command-line interface as you get comfortable with creating and configuring your style guide. You'll integrate Vale with your text editor in Chapter 4, Integrating Vale into Your Workflow, on page 53.

Once you've ensured Vale is available system-wide, you can configure it for a new project.

Setting Up Vale on Your Project

To get comfortable setting up Vale and using it to check content, you'll create a new project from scratch and add a file with some spelling issues.

Create a new project so you can experiment with Vale. First, create a directory called AwesomeCo and switch to that directory:

```
$ mkdir awesomeco
$ cd awesomeco
```

Within that directory, create a new Markdown file called README.md. In that file, add some content with intentionally misspelled words:

getting_started/setup/README.md
```
# AwesomeCo Code Repository

This is the github repo for AwesomeCo's company web site.
```

Now that you have at least one file to check, you can configure Vale with some rules. You need to add two things to your project's working directory to get Vale working: a .vale.ini configuration file and a directory to hold your custom style rules.

---

3.   https://snapcraft.io/
4.   https://github.com/errata-ai/vale/releases

First, create the .vale.ini file and add the following lines that tell Vale where to look for style rules and what kinds of rules you want to apply to the files in the project:

getting_started/setup/.vale.ini
```
StylesPath = styles
MinAlertLevel = suggestion

[*]
BasedOnStyles = Vale
```

The StylesPath entry specifies the directory that holds any style rules and vocabularies you'll add or create. In this example, you're using a styles directory in the same folder as the .vale.ini file. Don't create the folder yet, though; you'll let Vale do it for you.

The MinAlertLevel line tells Vale what severity level to use for alerts. There are three levels: error, warning, and suggestion, which is the default. You can define the severity level for each rule either when you define the rule or directly in the configuration file, which you'll explore shortly. The default MinAlertLevel is suggestion, but setting this explicitly in your configuration file makes clearer what you intend to report and also lets you change the setting later without having to look it up.

The [*] line targets all files in the project, regardless of the file type. Vale supports Markdown, HTML, and other popular formats. You can specify different rules based on file type, but to start, you'll treat every file the same.

The BasedOnStyles line tells Vale which style rules you want to use for the type of file you're scanning. Vale comes with a style each for spelling, checking terms, and checking repeated words. When you add more styles to your project, you'll add them to this line to enable them.

Now you can create the directory to hold your custom style definitions. The vale sync command downloads and updates premade styles that others have shared. You'll explore that in the next chapter, but it's also a quick way to have Vale create the folder you specified in your .vale.ini file. This ensures they match.

Run the vale sync command in your terminal:

```
$ vale sync
 SUCCESS  Synced 0 package(s) to '/Users/brianhogan/awesomeco/styles'.
```

You have a configuration file and a text file to scan. Use Vale to check the README.md file for errors, as shown on the next page:

```
$ vale README.md
README.md
 3:13  error  Did you really mean 'github'?  Vale.Spelling
 3:20  error  Did you really mean 'repo'?    Vale.Spelling
✖ 2 errors, 0 warnings and 0 suggestions in 1 file.
```

Vale found two spelling issues it classified as errors: github and repo. Errors cause Vale to return a nonzero exit code, which means that if this were part of a build process, it would break the build.

Controlling Error Levels

By default, spelling issues raise errors, which causes Vale to return a nonzero exit code. If you're using Vale as part of your continuous build system, like the one you'll set up in Using Vale with GitHub Actions, on page 72, any errors Vale raises will break your build.

When introducing Vale to an existing project, you probably won't want spelling errors to disrupt your team's workflow until you get them all fixed.

You can control the severity level Vale reports for each rule. In the .vale.ini file, add the following line at the end of the file to tell Vale that you want it to treat spelling issues as warnings instead of errors:

```
getting_started/setup/.vale.ini
StylesPath = styles
MinAlertLevel = suggestion

[*]
BasedOnStyles = Vale
Vale.Spelling = warning
```

Now run Vale on the README.md file again, and the two spelling errors are now warnings:

```
$ vale README.md
README.md
 3:13  warning  Did you really mean 'Github'?  Vale.Spelling
 3:20  warning  Did you really mean 'repo'?    Vale.Spelling
✖ 0 errors, 2 warnings and 0 suggestions in 1 file.
```

This time, Vale won't break any build process. But you should fix the issues anyway. Open README.md and change github to GitHub and repo to repository. Then run Vale again:

```
$ vale README.md
✓ 0 errors, 0 warnings and 0 suggestions in 1 file.
```

You might be wondering why Vale didn't flag AwesomeCo as a word. Vale tends to allow camel-case and Pascal-case words, which is why you were able to make the spelling error go away when you changed github to GitHub.

Detecting Repeated Words

Vale's built-in styles can detect repeated words. Add an extra the to the sentence in the README.md file to create a common scenario that even seasoned editors miss:

getting_started/multiple/README.md
```
# AwesomeCo Code Repository

This is the the GitHub repository for AwesomeCo's company web site.
```

Now run Vale against README.md again:

```
$ vale README.md

 README.md
 3:9  error    'the' is repeated!   Vale.Repetition

✖ 1 error, 0 warnings and 0 suggestions in 1 file.
```

Leave this error alone for now. You'll fix it later.

You've explored Vale with Markdown. Now try it with HTML.

Checking Multiple Formats

You've already configured Vale to apply its rules to all file types in your project, so test Vale out on an HTML file.

Create the file index.html in your project and add the following code to the file, complete with some misspelled words:

getting_started/multiple/index.html
```
<!DOCTYPE html>
<html lang="en-US">
  <head>
    <meta charset="utf-8">
    <title>AwesomeCo</title>
  </head>
  <body>
    <h1>Welcome To The The AwesomeCo Website</h1>

    <p>We enfuse your strategy with sinergy.</p>

  </body>
</html>
```

Save the file and run Vale on index.html:

```
$ vale index.html

index.html
8:20    error     'The' is repeated!               Vale.Repetition
10:11   warning   Did you really mean 'enfuse'?    Vale.Spelling
10:37   warning   Did you really mean 'sinergy'?   Vale.Spelling

✖ 1 error, 2 warnings and 0 suggestions in 1 file.
```

As expected, Vale found the two spelling errors and the repeated word. It ignored the HTML tags and found just the prose.

Check both files at the same time by passing each file as an argument:

```
$ vale index.html README.md

index.html
8:20    error     'The' is repeated!               Vale.Repetition
10:11   warning   Did you really mean 'enfuse'?    Vale.Spelling
10:37   warning   Did you really mean 'sinergy'?   Vale.Spelling

README.md
3:9     error     'the' is repeated!             Vale.Repetition
3:24    warning   Did you really mean 'repo'?    Vale.Spelling

✖ 2 errors, 3 warnings and 0 suggestions in 2 files.
```

You can also specify wildcards:

```
$ vale *.html *.md
```

You can also tell Vale to check a directory and let it sort through the files for you. Try the following command:

```
$ vale .
```

This time, you see an extra file in the results, your .vale.ini file:

```
.vale.ini
6:17 error   'Vale' is repeated!  Vale.Repetition

README.md
3:9     error     'the' is repeated!             Vale.Repetition
3:24    warning   Did you really mean 'repo'?    Vale.Spelling

index.html
8:20    error     'The' is repeated!               Vale.Repetition
10:11   warning   Did you really mean 'enfuse'?    Vale.Spelling
10:37   warning   Did you really mean 'sinergy'?   Vale.Spelling

✖ 3 errors, 3 warnings and 0 suggestions in 3 files.
```

This is an expected result; in your .vale.ini file, you specified that you want Vale to check all files, so that's what it's doing. Vale uses the patterns in .vale.ini to determine the rules to apply to the files you pass in as an argument.

You have to make sure Vale only checks the files you want it to scan, especially if you have lots of files in your repository. Vale is fast, but more files will slow you down, and you'll end up reading through lots of irrelevant error messages if you're not selective enough.

Open .vale.ini and change the pattern so it only looks for HTML and Markdown files:

getting_started/multiple/.vale.ini
```
StylesPath = styles
MinAlertLevel = suggestion

[*.{html,md}]
BasedOnStyles = Vale
Vale.Spelling = warning
```

Now, when you scan again, Vale only checks those specified files.

```
$ vale .

 index.html
 7:20   error     'The' is repeated!                Vale.Repetition
 9:11   warning   Did you really mean 'enfuse'?     Vale.Spelling
 9:37   warning   Did you really mean 'sinergy'?    Vale.Spelling

 README.md
 3:9    error     'the' is repeated!   Vale.Repetition

✖ 2 errors, 2 warnings and 0 suggestions in 2 files.
```

You can apply different sets of rules for different types of files using this approach. For example, you may want to check the spelling on the Markdown files but skip it on the HTML files because another process generated those pages, so you don't need to scan them. You can do that by changing how the rules apply.

Try it out. Change the .vale.ini file so the Vale style is active for HTML and Markdown files, but spell-checking is off for HTML files, and Markdown files get warnings instead of errors for spelling errors:

getting_started/folders/.vale.ini
```
StylesPath = styles
MinAlertLevel = suggestion

[*.{html,md}]
BasedOnStyles = Vale
```

```
[*.html]
Vale.Spelling = NO

[*.md]
Vale.Spelling = warning
```

Now, in your README.md file, change "repository" back to "repo" so Vale will flag it as a spelling mistake.

getting_started/folders/README.md
```
# AwesomeCo Code Repository

This is the the GitHub repo for AwesomeCo's company web site.
```

And yes, leave the extra "the" in the sentence.

Your Markdown and HTML files both have spelling issues. Run Vale against all files in the current directory:

```
$ vale .
```

This time, Vale ignored the spelling errors in HTML files but still found the repeated word, and it flagged the use of "repo" as a spelling error in the Markdown file:

```
index.html
8:20  error   'The' is repeated!  Vale.Repetition

README.md
3:9   error    'the' is repeated!             Vale.Repetition
3:24  warning  Did you really mean 'repo'?  Vale.Spelling
✖ 2 errors, 1 warning and 0 suggestions in 2 files.
```

Now you'll see how Vale finds configuration files and applies rules.

Getting Insight into Vale's Configuration

It's important to understand how Vale applies rules to files in your project because it impacts how you set up your rules in larger projects.

First, Vale's configuration file applies to everything inside the current directory, including subdirectories.

Create a new folder called website and move the index.html file into that folder:

```
$ mkdir website
$ mv index.html website/index.html
```

Now run Vale in the root directory of your project:

```
$ vale .
README.md
3:9    error    'the' is repeated!          Vale.Repetition
3:24   warning  Did you really mean 'repo'?  Vale.Spelling

website/index.html
8:20   error    'The' is repeated!  Vale.Repetition
```

Vale still found your HTML file and included it in the report. Vale looks in the current directory for a configuration file, and if it doesn't find one, it looks for one in the parent directory. If it works its way up the directory tree and still can't find one, it looks for a global configuration file.

The command vale ls-config prints out the applied Vale configuration in JSON format so you can get more insight into what's happening.

```
$ vale ls-config
```

The output contains a lot of information. Look for the RootINI and Paths sections at the end of the file:

```
...
  "RootINI": "/Users/brianhogan/vale/.vale.ini",
  "Paths": [
    "/Users/brianhogan/Library/Application Support/vale/styles",
    "/Users/brianhogan/vale/styles"
  ],
  "ConfigFiles": [
    "/Users/brianhogan/vale/code/.vale.ini"
  ],
}
```

The output shows that the .vale.ini file in the current directory is the root configuration file, and it matches the ConfigFiles entry as well. No other configuration files are active.

Switch to the website directory and run the vale ls-config command again, and the configuration shows the same:

```
$ cd website
$ vale ls-config
```

This time, Vale finds the configuration file in the parent directory:

```
...
  "RootINI": "/Users/brianhogan/books/vale/code/.vale.ini",
  "Paths": [
    "/Users/brianhogan/Library/Application Support/vale/styles",
    "/Users/brianhogan/books/vale/code/styles"
  ],
```

```
  "ConfigFiles": [
    "/Users/brianhogan/books/vale/code/.vale.ini"
  ],
```

Switch back to the root of your project again and run vale ls-config again. This time, look at the RuleToLevel section:

```
{
  "RuleToLevel": {
    "Vale.Spelling": "warning"
  },
  ...
```

From this output, you can see that the current Vale configuration sets the error level for the Vale.Spelling rule to warning. You can't specify different warning levels for different file types. When Vale combines the rules, the error level for each rule gets set globally.

The output also shows how Vale will apply specific rules. Look at the SBaseStyles and SChecks sections, and you'll recognize your rules:

```
"SBaseStyles": {
  "*.{html,md}": [
    "Vale"
  ]
},
"SChecks": {
  "*.html": {
    "Vale.Spelling": false
  },
  "*.md": {
    "Vale.Spelling": true
  },
  "*.{html,md}": {}
},
```

As you add more rules, file types, and settings, you can use vale ls-config to inspect and debug your configuration.

Depending on your needs, you may want to create content-specific configuration files.

Using Different Configuration Files

Sometimes you'll have to use different rules for different types of content, like one style for blog posts and another for your technical documentation. They both might be the same kind of content type, so you can't separate the rules based on the file type. To address for this, use different configuration files for each type.

In the website folder, create a new .vale.ini file that only looks at HTML files and ensures that spelling errors cause errors:

```
getting_started/folders/website/.vale.ini
StylesPath = ../styles
MinAlertLevel = suggestion

[*.html]
BasedOnStyles = Vale
Vale.Spelling = error
```

Notice that in this configuration file, the StylesPath points to the styles directory in the parent directory rather than in the current directory.

Now change to the website directory and run Vale from there:

```
$ cd website
$ vale .
```

This time, the spelling errors display, and they're flagged as errors:

```
 index.html
 8:20    error   'The' is repeated!             Vale.Repetition
 10:11   error   Did you really mean 'enfuse'?  Vale.Spelling
 10:37   error   Did you really mean 'sinergy'? Vale.Spelling

✖ 3 errors, 0 warnings and 0 suggestions in 1 file.
```

Remember, unless you specify a configuration file, Vale will look for one in the current directory and then start looking at parent directories. Using directory-specific Vale configuration files is a good way to create section-specific configurations in a content monorepo, where each team can control the rules for their section of content.

When you run Vale from the same directory as a configuration file, you don't need to explicitly tell Vale which configuration file to use. But using the --config option, you can be explicit.

Go back to the parent directory that holds your README.md file:

```
$ cd ..
```

Now run Vale against all files, but pass in the configuration file in the website folder:

```
$ vale --config "website/.vale.ini" .
```

```
 website/index.html
 8:20    error   'The' is repeated!             Vale.Repetition
 10:11   error   Did you really mean 'enfuse'?  Vale.Spelling
 10:37   error   Did you really mean 'sinergy'? Vale.Spelling

✖ 3 errors, 0 warnings and 0 suggestions in 1 file.
```

This gives the same result as before, since you specifically told Vale which configuration file to use rather than letting it figure it out.

You now have a couple of ways to organize your configuration files and use them when linting content.

Overriding Files to Scan

Right now, you have a .vale.ini file in your project root and a website folder with its own .vale.ini file. If you ran vale . in the root of your project, it would recursively scan all the files in your project and use the Vale configuration in the current directory. That would mean the Vale rules in the website directory wouldn't get applied to files in the website directory.

The --glob option lets you specify a pattern that overrides the file paths in the Vale.ini configuration. You can use this pattern to specify files to include, but you can also use it to exclude files by using the ! character in front of the glob pattern.

Run the following command to tell Vale to scan all files in the project directory and subdirectories, except for anything in the website folder or its child folders:

```
$ vale --glob='!website/**/*' .
```

The resulting output only shows the README.md file:

```
README.md
 3:9    error    'the' is repeated!             Vale.Repetition
 3:24   warning  Did you really mean 'repo'?    Vale.Spelling

✖ 1 error, 1 warning and 0 suggestions in 1 file.
```

The --glob flag gives you another way to control how Vale scans your content. It takes precedence over the paths in the .vale.ini file.

Your Turn

Before moving on, complete these challenges to ensure you're comfortable with Vale's command-line interface:

1. Look at the options presented by vale --help and get the output of the report in JSON format. What's represented in the output? When would you use this kind of output?
2. Look at the options and see if you can get Vale to only show errors, not suggestions or warnings. Hint: it's not the --filter option; you'll explore that in the next chapter.

3. Correct the errors in your HTML and Markdown files so you get no errors when you scan.

Wrapping Up

You've configured Vale to check two different file types for spelling issues, and you looked at how Vale's configuration works at a high level. You can start integrating Vale with your projects right now and get some immediate benefits. Next, you'll use Vale to download and run rules that enforce some popular style guides.

CHAPTER 2

Using Existing Styles

Building a style guide from scratch is no small task. That's why lots of content publishers adopt a well-known one and adapt it. Microsoft, Google, Grafana, RedHat, and GitLab have all published their style guides either on their public websites or as part of their documentation repositories on GitHub. Best of all, these companies use Vale to enforce those styles, so you can start with these, peek at their rules, and mix and match to find a good foundation.

That's what you'll do in this chapter. You'll start with a single style. Then you'll bring in another style and manage conflicting rules. Then you'll look at other styles that make content more inclusive and readable. To do that, you'll need some content to work with.

In the book's companion files, you'll find a directory called cli-tips, which is a collection of tutorials for some command-line Linux tools. This content is in rough shape, but you'll use Vale to work out the inconsistencies. You'll spend this chapter and the next building a set of rules for your own content. In this chapter you'll use a handful of existing style rules.

Visit the book's website[1] and download and extract the companion files for the book. Open a new Terminal and navigate to the cli-tips folder. You'll see five files:

```
$ cd cli-tips
$ ls
awk.md   date.md   find.md   grep.md   sed.md
```

Look at each file and you'll see that there are some quality concerns. It's time to get to work fixing things up.

---

1. https://thevalebook.com

Downloading Existing Styles with Vale

The Vale maintainer implemented the Microsoft and Google style guides and made them available on GitHub. You could copy them into your project, but Vale has a packaging system that lets you download styles and configurations and keep them in sync.

The vale sync command not only creates the directory for your styles, but it also downloads and updates any linked style rules you specify. Try it out by creating a new .vale.ini file in the cli-tips folder with the following contents:

```
use_existing_styles/single/.vale.ini
StylesPath = styles
MinAlertLevel = suggestion
Packages = Google

[*.{md}]
BasedOnStyles = Vale, Google
```

This configuration uses the Packages field. Packages let people share Vale styles with other Vale users. A package is a ZIP file that contains its own .vale.ini file, style rules, or both. If you see the package you want to use listed in Vale's Package Explorer,[2] you can use the package name, and Vale will automatically download it when you run vale sync. You can also provide a URL to a package someone shared, a path to a local ZIP file you downloaded, or a path to a local package you created.

Save the file and run vale sync. The Google style guide downloads.

```
$ vale sync
Syncing Google [1/1] ████████████████████████████  100% | 0s
 SUCCESS  Synced 1 package(s) to '/Users/brianhogan/cli-tips/styles'.
```

Now run Vale on the awk.md file:

```
$ vale awk.md

awk.md
 2:30   suggestion   Spell out 'AWK', if it's        Google.Acronyms
                     unfamiliar to the audience.
 6:1    suggestion   Spell out 'AWK', if it's        Google.Acronyms
                     unfamiliar to the audience.
 6:76   error        Did you really mean 'thats'?    Vale.Spelling
 8:4    warning      'Basic Structure' should use    Google.Headings
                     sentence-style capitalization.
 16:1   warning      Try to avoid using              Google.We
                     first-person plural like
                     'Let's'.
```

---

2. https://vale.sh/explorer

52:4	warning	'doing Calculations' should use sentence-style capitalization.	Google.Headings
54:1	warning	Try to avoid using first-person plural like 'We'.	Google.We
54:35	warning	Try to avoid using first-person plural like 'Let's'.	Google.We
60:11	error	Did you really mean 'xamples'?	Vale.Spelling
72:1	suggestion	Spell out 'AWK', if it's unfamiliar to the audience.	Google.Acronyms
72:80	error	Don't use exclamation points in text.	Google.Exclamation

✖ 3 errors, 5 warnings and 3 suggestions in 1 file.

Vale's output makes it clear that this content isn't aligned with Google's style. Now you have a road map of what to clean up. But don't make any changes yet. First, take a look at some of the definitions.

Exploring the Style

The styles folder in your project now has a new Google folder. That folder contains files in YAML format that define the rules for the style.

When you scanned the awk.md file, the last suggestion you saw was this one:

72:80	error	Don't use exclamation points in text.	Google.Exclamation

Look in styles/Google and you'll see a file called Exclamation.yml that contains the rule. Open the file in your editor to see the rule's definition:

```
use_existing_styles/single/styles/Google/Exclamation.yml
extends: existence
message: "Don't use exclamation points in text."
link: "https://developers.google.com/style/exclamation-points"
nonword: true
level: error
action:
  name: edit
  params:
    - trim_right
    - "!"
tokens:
  - '\w+!(?:\s|$)'
```

The rule includes the rule type (or "check"), a message, a link to the rule in the style guide, and most importantly, some kind of regular expression to match. In the next chapter, you'll become more familiar with these definitions as you create your own rules from scratch.

Some Vale rules include an action section that text editors and other tools can use to correct the errors Vale finds. In this case, the definition instructs the tool to remove the exclamation point. It's up to each editor or tool to handle the actual implementation. You'll explore using Vale with editors in Seeing Errors as You Work, on page 53.

Changing a Rule's Error Level

As you saw in Controlling Error Levels, on page 4, you can change a rule's severity in the .vale.ini file. The Google package you added to your project has many rules split into individual YAML files. You can change the severity of each rule by specifying it using the rule name listed in the output.

For example, the rule about exclamation points has the name Google.Exclamation:

```
72:80  error       Don't use exclamation points    Google.Exclamation
                   in text.
```

This rule raises an error rather than a warning or suggestion. While you could directly edit the YAML file, doing so means you have to track changes and reapply them if you ever update the rules. Upstream packages do change as companies evolve their styles.

Instead, use the .vale.ini file to raise warnings instead of errors:

```
use_existing_styles/single/.vale.ini
[*.{md}]
BasedOnStyles = Vale, Google
Google.Exclamation = warning
```

With the change in place, run Vale against the awk.md file again:

```
$ vale awk.md
```

The output shows warnings instead of errors this time:

```
...
 72:70  warning     Don't use exclamation points    Google.Exclamation
                    in text.
```

You can control the severity level for errors, or even turn individual rules on and off. That'll come in handy next, as you incorporate another style guide into your project.

Adding a Complementary Style

Sometimes an existing style guide doesn't catch all the cases you want to cover, so you may use another style guide as a fallback. You're not limited to using a single package in your Vale configuration. You can bring in more packages and mix and match the rules to get the results you're looking for.

Vale offers an implementation of the Microsoft style guide through its packaging system, which is a good fallback for the Google style guide. Open your .vale.ini file and change it so it includes the Microsoft style guide. Add Microsoft to the Packages key, and then add it to the BasedOnStyles key in your Markdown section:

```
use_existing_styles/multiple/.vale.ini
StylesPath = styles
MinAlertLevel = suggestion
Packages = Google, Microsoft

[*.{md}]
BasedOnStyles = Vale, Google, Microsoft
Google.Exclamation = warning
```

Run vale sync to download the new package:

```
$ vale sync
Syncing Microsoft [2/2] ██████████████████████████████ 100% | 1s
 SUCCESS  Synced 2 package(s) to '/Users/brianhogan/cli-tips/styles'.
```

Now run Vale against the awk.md file again. You'll see more errors, and some issues show up twice:

```
$ vale awk.md
awk.md
 2:30   suggestion  'AWK' has no definition.         Microsoft.Acronyms
 2:30   suggestion  Spell out 'AWK', if it's         Google.Acronyms
                    unfamiliar to the audience.
 6:1    suggestion  Spell out 'AWK', if it's         Google.Acronyms
                    unfamiliar to the audience.
 6:1    suggestion  'AWK' has no definition.         Microsoft.Acronyms
 6:11   warning     Remove 'extremely' if it's not   Microsoft.Adverbs
                    important to the meaning of
                    the statement.
 6:76   error       Did you really mean 'thats'?     Vale.Spelling
 8:4    warning     'Basic Structure' should use     Google.Headings
                    sentence-style capitalization.
 8:4    suggestion  'Basic Structure' should use     Microsoft.Headings
                    sentence-style capitalization.
 16:1   warning     Try to avoid using               Google.We
                    first-person plural like
                    'Let's'.
```

16:1	warning	Try to avoid using first-person plural like 'Let's'.	Microsoft.We
52:4	suggestion	'doing Calculations' should use sentence-style capitalization.	Microsoft.Headings
52:4	warning	'doing Calculations' should use sentence-style capitalization.	Google.Headings
54:1	warning	Try to avoid using first-person plural like 'We'.	Google.We
54:1	warning	Try to avoid using first-person plural like 'We'.	Microsoft.We
54:35	warning	Try to avoid using first-person plural like 'Let's'.	Microsoft.We
54:35	warning	Try to avoid using first-person plural like 'Let's'.	Google.We
60:11	error	Did you really mean 'xamples'?	Vale.Spelling
72:1	suggestion	'AWK' has no definition.	Microsoft.Acronyms
72:1	suggestion	Spell out 'AWK', if it's unfamiliar to the audience.	Google.Acronyms
72:80	warning	Don't use exclamation points in text.	Google.Exclamation

✖ 2 errors, 10 warnings and 8 suggestions in 1 file.

Both styles don't want you to use first-person plurals, both want you to spell out acronyms, and both want sentence-style headings. The Microsoft style guide flags adverbs like "extremely," which is something a lot of technical writers like to remove from writing. To get the best of both guides without repeating errors, turn off the duplicate rules in the Microsoft guide.

Modify the .vale.ini file to remove the duplicate rules:

```
use_existing_styles/multiple/.vale.ini
[*.{md}]
BasedOnStyles = Vale, Google, Microsoft
Google.Exclamation = warning
➤ # Turn off duplicate rules
➤ Microsoft.Acronyms = NO
➤ Microsoft.We = NO
➤ Microsoft.Headings = NO
```

Note that when you turn off a rule, case matters. The casing here matches the casing of the rule in the Vale output.

Run Vale again on awk.md and you'll see fewer results in the report, as the rules you excluded no longer apply:

```
$ vale awk.md
```

```
awk.md
 2:30    suggestion   Spell out 'AWK', if it's        Google.Acronyms
                      unfamiliar to the audience.
 6:1     suggestion   Spell out 'AWK', if it's        Google.Acronyms
                      unfamiliar to the audience.
 6:11    warning      Remove 'extremely' if it's not  Microsoft.Adverbs
                      important to the meaning of
                      the statement.
 6:76    error        Did you really mean 'thats'?    Vale.Spelling
 8:4     suggestion   'Basic Structure' should use    Microsoft.Headings
                      sentence-style capitalization.
 8:4     warning      'Basic Structure' should use    Google.Headings
                      sentence-style capitalization.
16:1     warning      Try to avoid using              Google.We
                      first-person plural like
                      'Let's'.
52:4     suggestion   'doing Calculations'            Microsoft.Headings
                      should use sentence-style
                      capitalization.
52:4     warning      'doing Calculations'            Google.Headings
                      should use sentence-style
                      capitalization.
54:1     warning      Try to avoid using              Google.We
                      first-person plural like 'We'.
54:35    warning      Try to avoid using              Google.We
                      first-person plural like
                      'Let's'.
60:11    error        Did you really mean 'xamples'?  Vale.Spelling
72:1     suggestion   Spell out 'AWK', if it's        Google.Acronyms
                      unfamiliar to the audience.
72:80    warning      Don't use exclamation points    Google.Exclamation
                      in text.

✖ 2 errors, 7 warnings and 5 suggestions in 1 file.
```

You can build a comprehensive linter solely by collecting rules from existing styles and turning off duplicate rules or rules you want to ignore.

Next, you'll make your text more inclusive and accessible.

Checking for Readability, Profanity, and Inconsiderate Writing

Vale's author maintains two other packages that can help you wrangle the language you use in your content, and these are especially helpful if you're working with content from external contributors. The Readability package helps you check your content for complexity, and the Alex package,[3] a Vale-compatible version of Alex,[4] a tool for testing insensitive writing.

Testing Readability

Technical writing may have lots of jargon, but you still want the content to be approachable by a wider audience. Making content more approachable will also help if you have to translate your content into other languages. The Readability package runs several algorithms against the content to see how readable it is.

Add the Readability package to your project. Open .vale.ini and add it to both the Packages and BasedOnStyles keys:

```
use_existing_styles/readability/.vale.ini
StylesPath = styles
MinAlertLevel = suggestion

Packages = Google, Microsoft, Readability

[*.{md}]
BasedOnStyles = Vale, Google, Microsoft, Readability
Google.Exclamation = warning
Microsoft.Acronyms = NO
Microsoft.We = NO
Microsoft.Headings = NO
```

Now run vale sync to download the new Readability package and update any other packages that might have changed:

```
$ vale sync
Syncing Readability [3/3]  ████████████████████████  100% | 2s
 SUCCESS  Synced 3 package(s) to '/Users/brianhogan/cli-tips/styles'.
```

Now run Vale again, and you'll find new readability scores at the top of the output:

```
awk.md
 1:1  warning    Try to keep the SMOG grade      Readability.SMOG
                  (10.29) below 10.
 1:1  warning    Try to keep the LIX score       Readability.LIX
                  (38.06) below 35.
 1:1  warning    Try to keep the Coleman–Liau    Readability.ColemanLiau
                  Index grade (11.74) below 9.
 1:1  warning    Try to keep the Flesch reading  Readability.FleschReadingEase
                  ease score (52.24) above 70.
 ...
```

---

3. https://github.com/errata-ai/alex
4. https://github.com/get-alex/alex

Vale reports these first because they operate on the entire file rather than by matching a specific line number and character position, so you'll find them at the top of the list.

The Readability package includes several different readability tests. For example, the Flesch Reading Ease test uses this formula:

```
206.835 - (1.015 * (words / sentences)) - (84.6 * (syllables / words))
```

A higher score means the content is easier to read, whereas a lower score means it's more challenging. A score between 60 and 70 targets the 8th-grade reading level, which is a good target for general audience writing. In this case, the score comes in around 52, which is getting close to a college reading level.[5]

Look at the YAML files in the styles/Readability. You'll find a link in each one that explains the test and how to measure each result.

Improving readability isn't the only way to make your writing more approachable.

Finding Inconsiderate Words

Alex[6] is a popular tool that finds polarizing language. Whether you want to remove profanity from user-generated content or avoid accidental use of insensitive phrases like "grandfathered" or "blacklist" in your writing, Alex has you covered. Vale provides a compatible implementation of the Alex guidelines so you can add it to your project without adding another tool or maintaining your own word list.

Open .vale.ini and add the alex package to Packages and BasedOnStyles:

```
use_existing_styles/alex/.vale.ini
StylesPath = styles
MinAlertLevel = suggestion
Packages = Google, Microsoft, Readability, alex

[*.md]
BasedOnStyles = Vale, Google, Microsoft, Readability, alex
Google.Exclamation = warning
Microsoft.Acronyms = NO
Microsoft.We = NO
Microsoft.Headings = NO
```

---

5. https://en.wikipedia.org/wiki/Flesch%E2%80%93Kincaid_readability_tests
6. https://github.com/get-alex/alex

Save the file and run vale sync to download the alex package:

```
$ vale sync
Syncing alex [4/4] ████████████████████████████████████ 100% | 1s
  SUCCESS  Synced 4 package(s) to '/Users/brianhogan/cli-tips/styles'.
```

Run Vale against awk.md again, and you'll see two new errors mixed in with the rest:

```
$ vale awk.md
...
 16:17   error       Using 'simple' may come across              alex.Condescending
                     as condescending.
...
 72:11   warning     When referring to a person, consider using  alex.Ablist
                     'incredibly' instead of 'insanely'.
```

You'll find these packages and rules aren't perfect. They're not context-aware, so they may raise false positives. For example, the Alex package thinks the word invalid is potentially ableist. But if you've used it to mean "not valid," Vale won't know that.

You should treat Alex as writing advice rather than a strict set of rules. Alter the error severity and turn off rules you don't think apply or don't make sense for your situation.

You've added four packages to your project. It's starting to get overwhelming, and you've only scanned one file.

Filtering Results

You'll find situations where you want to run a subset of rules. For example, if you wanted to check the readability scores of all your content right now, you'd have a lot of results to read through that aren't related to the readability scores. You can filter the rules Vale uses by using the --filter CLI option and an expression. You can filter on rule name, error level, scope, the link field in a rule, or even the rule's type, like substitutions or metrics rules. You write these rules using the expr expression language.[7]

For example, you can show rules where the message contains the word "capitalization":

```
.Message contains "capitalization"
```

Or you can filter out suggestions:

```
.Level != "suggestion"
```

---

7.   https://expr-lang.org/docs/language-definition

Or you can create a filter that only shows readability scores for your content. To do so, you'd use this expression:

```
.Name startsWith "Readability"
```

To run these filters, use the --filter flag:

```
$ vale --filter='.Name startsWith "Readability"' awk.md
```

```
awk.md
 1:1  warning  Try to keep the Coleman—Liau     Readability.ColemanLiau
                Index grade (11.74) below 9.
 1:1  warning  Try to keep the SMOG grade       Readability.SMOG
                (10.29) below 10.
 1:1  warning  Try to keep the LIX score        Readability.LIX
                (38.06) below 35.
 1:1  warning  Try to keep the Flesch reading   Readability.FleschReadingEase
                ease score (52.24) above 70.

✖ 0 errors, 4 warnings and 0 suggestions in 1 file.
```

While you can write ad-hoc queries, if you plan to reuse them, you can save the expressions for later use to save some typing. A readability filter is a great candidate for a saved filter.

Vale expects filters to live in the config/filters folder within the styles directory. Create this directory:

```
$ mkdir -p styles/config/filters
```

Then create a new file called readability.exp in that folder that contains the following expression:

```
use_existing_styles/alex/styles/config/filters/readability.exp
.Name startsWith "Readability"
```

Now run Vale and pass the filename of the filter to the --filter argument:

```
$ vale --filter=readability.exp awk.md
```

```
awk.md
 1:1  warning  Try to keep the Coleman—Liau     Readability.ColemanLiau
                Index grade (11.74) below 9.
 1:1  warning  Try to keep the SMOG grade       Readability.SMOG
                (10.29) below 10.
 1:1  warning  Try to keep the LIX score        Readability.LIX
                (38.06) below 35.
 1:1  warning  Try to keep the Flesch reading   Readability.FleschReadingEase
                ease score (52.24) above 70.

✖ 0 errors, 4 warnings and 0 suggestions in 1 file.
```

Vale doesn't report the rest of the rules. This is how you can run specific rules without maintaining a separate .vale.ini file.

Your Turn

Try these challenges before moving on to the next chapter:

1. Scan the rest of the Markdown files in the cli-tips directory and identify other duplicate rules. Add those to the list of duplicate rules.
2. Look at the Google and Microsoft acronyms list. Is one more comprehensive than the other? Which one would you use?
3. In the Readability package, you'll find a Vale rule for each type of readability metric. You'll also find a link in each one to an explanation. Read through the metrics and turn off the metrics you don't think apply to your work. If you turn off a rule through your .vale.ini file, remember that the rule name is case-sensitive; it matches the name of the rule on the Vale output report.
4. Modify the readability filter you created so it only runs the Flesch Reading Ease test, and run it against all your files.
5. Create a filter that only shows errors. Then adjust the error level for those rules in your .vale.ini file so they are warnings.

Wrapping Up

You can make massive improvements to your content's accessibility, inclusiveness, and consistency by using existing styles against your content. You can take what you need and discard what you don't, and in no time, you'll have a road map of what you need to change.

Style rules aren't one-size-fits-all. While you can turn individual YAML rule files off, you can't remove certain words from a word list. And you may have your own key terms. Next, you'll write your own rules and build your own style guide.

CHAPTER 3

Building Your Own Style

So far, you checked documents using the existing style packages you installed through Vale. While those rules cover a lot, you probably have your own style rules you want to enforce that are unique to the kind of content you're producing.

Rather than relying solely on existing style packages, most organizations create their own style guide, writing new rules when necessary and using rules from other places when it makes sense. That's what you'll do now. You'll start by creating a new style and writing some rules from scratch.

Creating a New Style

When you ran vale sync, Vale downloaded the packages into folders within the styles directory. To add your own style, create your own folder in that directory with the name of the style. Add a new "AwesomeCo" style by creating the styles/AwesomeCo folder:

```
$ mkdir styles/AwesomeCo
```

The name you give the folder is case-sensitive, because it becomes the rule's prefix you see in reports and the prefix you use when turning rules on and off in your configuration.

Now you need some rules to enforce your style guide. The first rule you'll make is one that looks for the word "utilize" and suggests "use" as the alternative. In most cases, "utilize" isn't necessary. It's also a good example for creating a basic Vale rule.

Create a new YAML file in styles/AwesomeCo called BadWords.yml. Add the following to the file:

```
build_your_style/rules/styles/AwesomeCo/BadWords.yml
extends: substitution
message: "Use '%s' instead of '%s'."
ignorecase: true
level: error
action:
  name: replace
swap:
  - utilize: use
```

Use .yml, Not .yaml

Vale only allows the .yml file extension for rules. If you use the .yaml extension, Vale won't see your rules.

The existence key tells Vale what kind of rule you're creating. Each Vale rule you write extends one of Vale's built-in checks. The substitution check lets you build a rule that looks for a word and suggests another word. You'll explore other Vale checks in this chapter.

The message key lets you specify the message Vale displays in the error report. The substitution check returns the word it found and the suggested substitution, and you can insert those words into the message using string replacement. The first %s references the word Vale flagged, and the second references the suggested alternative.

ignorecase tells Vale that this rule isn't case-sensitive, so it'll flag utilize no matter how it's capitalized in the content. In rules where you want to enforce certain product names or jargon, you'll set this to false.

error lets you set the rule's default error level. In this case, you're flagging utilize as a word that the writer needs to fix. You can override this in the .vale.ini file.

In some rules, you can define an "action" that tells tools that use Vale how to automatically fix the errors it finds. In this case, you're defining a replace action, which tells the tool to use the suggested word. Vale doesn't fix things on its own, but various text editors can. You'll configure Vale with your editor in Seeing Errors as You Work, on page 53.

Finally, the swap field is where you specify the words or phrases to look for, followed by their alternatives. Each entry gets its own line, starting with a dash. The word or phrase comes first, followed by a colon, and then the suggested replacement.

Save the rule. Now edit the .vale.ini file in the project to use your AwesomeCo style and no other styles for Markdown files:

```
build_your_style/rules/.vale.ini
StylesPath = styles
MinAlertLevel = suggestion
Packages = Google, Microsoft, Readability, alex

[*.md]
BasedOnStyles = AwesomeCo
```

Run Vale against all files now, and you'll see one violation:

```
$ vale .

 sed.md
 6:94   error   Use 'use' instead of        AwesomeCo.BadWords
                'utilize'.

✖ 1 error, 0 warnings and 0 suggestions in 6 files.
```

This version only catches "utilize", either capitalized or uncapitalized. It won't catch variations like "utilized," "utilizes," and "utilizing." To add those, add substitution entries for those too:

```
build_your_style/rules/styles/AwesomeCo/BadWords.yml
extends: substitution
message: "Use '%s' instead of '%s'."
ignorecase: true
level: error
action:
  name: replace
swap:
  - utilize: use
  - utilized: used
  - utilizes: uses
  - utilizing: using
```

You've written a basic rule that not only flags a word, but offers a possible substitution. Next, you'll look at the existence check, which you'll use when you need to flag something that might not have a quick substitution.

Ensure Certain Phrases Don't Exist

Vale is great at helping you remove filler words. Sometimes you don't want to substitute a word or phrase; sometimes you want to make sure it's not included at all. That's what the existence check is for.

The phrases "learn to" or "learn how to" often add noise in technical content. When you write "Learn how to search for text with grep," you're inadvertently placing emphasis on the learning process rather than the skill you're developing.

Reducing the sentence to "Search for text with grep" is stronger and emphasizes the verb "search" instead.

Add a rule that enforces this style. Create the file styles/AwesomeCo/Learn.yml:

```
build_your_style/rules/styles/AwesomeCo/Learn.yml
extends: existence
message: "Consider avoiding '%s' as it focuses on learning rather than doing."
ignorecase: true
level: warning
action:
  name: remove
tokens:
  - 'learn(?: how)? to'
```

In an existence rule, you use the tokens key to specify words, phrases, or regular expressions you want Vale to flag.

Now run Vale against your files, and you'll see two new warnings in addition to your previous error:

```
$ vale .

 grep.md
 24:1   warning   Consider avoiding 'Learn how      AwesomeCo.Learn
                  to' as it focuses on learning
                  rather than doing.



 sed.md
 3:10   warning   Consider avoiding 'Learn to'      AwesomeCo.Learn
                  as it focuses on learning
                  rather than doing.
 6:94   error     Use 'use' instead of             AwesomeCo.BadWords
                  'utilize'.

✖ 1 error, 2 warnings and 0 suggestions in 6 files.
```

Create rules based on the suggestion check when you want to suggest alternate wording, and use existence when you want to check if something exists that shouldn't be there.

Dealing with Nonword Characters

When tokens contain punctuation, you have to change how the existence rule operates.

Sometimes tutorials end with some variation of "Congratulations! You've completed the tutorial." While the author may be well-meaning, readers can find this patronizing. Add a rule that looks for the word congratulations, either with or without an exclamation point at the end.

Create the file styles/AwesomeCo/Patronizing.yml that includes the following rule:

```
build_your_style/rules/styles/AwesomeCo/Patronizing.yml
extends: existence
message: "Avoid patronizing phrases like '%s'."
ignorecase: true
level: warning
action:
  name: remove
nonword: true
tokens:
  - 'congratulations!?'
```

By default, existence rules convert tokens into a regular expression that uses word boundaries. In this case, it converts the token congrationatls!? to this:

```
(?i)(?m)\b(?:congratulations!?)\b
```

This results in the rule matching only congratulations, but not including the exclamation point in the capture group. That may not seem like a big deal at first, but the %s in the message uses the captured text in the output. And Vale uses it for start and end positions, which also affects the action specified.

When you use nonword: true, then Vale won't use word boundaries, so the regex captures the exclamation point.

Run Vale against all your files, and you'll see the new error:

```
$ vale .
 grep.md
 24:1  warning  Consider avoiding 'Learn how    AwesomeCo.Learn
                to' as it focuses on learning
                rather than doing.
 76:1  warning  Avoid patronizing phrases like  AwesomeCo.Patronizing
                'Congratulations!'.

 sed.md
 3:10  warning  Consider avoiding 'Learn to'    AwesomeCo.Learn
                as it focuses on learning
                rather than doing.
 6:94  error    Use 'use' instead of            AwesomeCo.BadWords
                'utilize'.

✖ 1 error, 3 warnings and 0 suggestions in 6 files.
```

Next, you'll look for occurrences of words, phrases, or characters in the document.

Ensuring Something Exists in the Document

The existence check helps you find things in a document that shouldn't be there, but you'll also want to ensure that some phrase, word, or heading occurs one or more times in text. Depending on what you're looking to do, you have two options for this: occurrence or conditional.

The occurrence check lets you specify the number of times you want a word, phrase, or character to occur in a given scope. It does this by counting the number of times the token value occurs within the scope. The conditional check lets you check that if one token exists, another does as well.

Ensuring Something Occurs Within a Range

In Testing Readability, on page 22, you added the Readability package to your style, which scored your content based on several readability formulas. Sentence and paragraph lengths can also have an effect on readability, and you can use the occurrence check to count words in sentences or paragraphs.

Add the styles/AwesomeCo/ParagraphLength.yml file and the following rule that checks paragraphs to ensure they are between 100 and 200 words:

```
build_your_style/rules/styles/AwesomeCo/ParagraphLength.yml
extends: occurrence
message: "Try to keep paragraphs between 100 and 200 words. (%s)."
scope: paragraph
level: suggestion
min: 100
max: 200
token: \b(\w+)\b
```

This rule looks at each paragraph on its own and applies the regular expression in the token field. Since the expression looks for words, each word is a match. As long as the number of matches in that scope is between the min and max values, there won't be any errors raised.

None of the content you have in the cli-tips folder has large paragraphs, but 100 to 200 words is a good metric for general-purpose writing. Run this rule against your files to ensure it works, and then turn it off in your .vale.ini configuration:

```
build_your_style/rules/.vale.ini
[*.md]
BasedOnStyles = AwesomeCo
AwesomeCo.ParagraphLength = NO
```

You can always turn the rule on later or use it in a different configuration.

When working with the occurrence check and a specific scope, like heading or paragraph, Vale looks for matches within each scope, not in the whole document. When you specify scope: paragraph, the rule applies to each paragraph, not across all paragraphs. Think of it as looping over each paragraph and executing the check. It's unaware of anything outside of that scope during the check. This means that you can use it to test for punctuation in a heading, for example, but you can't use it to ensure a specific heading occurs once in your document. You'll have to use a different scope for that, which you'll do later in Writing Custom Rules with Scripts, on page 58.

Ensuring Something Exists Based on Something Else

The conditional check lets you look for one pattern based on another. The most common use for this is the one Vale outlines in its own documentation: checking to see if an acronym or abbreviation has a definition first.

Add this rule to your project. Create styles/AwesomeCo/FirstUse.yml with the following rule that looks for an acronym of three to five characters and then looks to see if that same acronym exists wrapped in parentheses:

```
build_your_style/rules/styles/AwesomeCo/FirstUse.yml
extends: conditional
message: "Define acronyms and abbreviations on first use. ('%s')"
ignorecase: false
level: suggestion
first: '\b([A-Z]{3,5}s?)\b'
second: '\(([A-Z]{3,5}s?)\)'
exceptions:
  - HTML
  - JSON
  - URL
  - ZIP
  - YAML
```

This rule assumes that you'll define terms with the spelled-out version, followed by the acronym or abbreviation in parentheses.

This rule uses the exceptions key to specify some tokens that won't trigger the rule. In this case, you don't need to define HTML and JSON first. If you have other known acronyms or abbreviations, you can add them as exceptions here as well.

Run this rule against your files, and you'll see a few new suggestions, including these in the AWK file:

```
$ vale .

...

awk.md
  2:30     suggestion   Define acronyms and           AwesomeCo.FirstUse
                        abbreviations on first use.
                        ('AWK')
  6:1      suggestion   Define acronyms and           AwesomeCo.FirstUse
                        abbreviations on first use.
                        ('AWK')
  6:229    suggestion   Define acronyms and           AwesomeCo.FirstUse
                        abbreviations on first use.
                        ('CSV')
  38:21    suggestion   Define acronyms and           AwesomeCo.FirstUse
                        abbreviations on first use.
                        ('CSV')
  72:1     suggestion   Define acronyms and           AwesomeCo.FirstUse
                        abbreviations on first use.
                        ('AWK')

...
```

The result flags AWK and CSV. CSV is as common as HTML and JSON, so you could add it to the exceptions in your rule, or you could alter the content to spell it out first. As for AWK, that's a reference to the AWK language[1] and the associated CLI tool, awk. You'll have to determine how you'll treat this word in your content. You could create a substitution rule for this to let people know it should be lowercase, or you could add an exception for it in the Vale rule.

Leave these alone for now. In Overriding Styles with Accepted Terms, on page 46, you'll explore another approach to handling terms.

Ensuring Something Exists in the Entire Document

Vale is markup-aware, meaning that it does its best to extract the text from Markdown, HTML, and other formats, while ignoring things like YAML front matter, inline code, and code blocks, as well as HTML tags and scripts. You'll find times when you want to use Vale to look at certain things that it skips.

For example, you may want to make sure that each of your tutorials has a Conclusion heading. You might think you can create a rule based on the occurrence check for that, but since Vale strips the markup out, you won't be able to identify headings. If you changed the scope to look only at headings, Vale would check each heading one at a time rather than look at all headings in the document.

---

1. https://en.wikipedia.org/wiki/Regular_expression

To solve this, use scope: raw. This instructs Vale to look at the entire raw document without processing it first. This means your rule sees every character, including code, markup, and special characters.

Test it out. Create a new file called styles/AwesomeCo/Conclusion.yml with this content:

```
build_your_style/rules/styles/AwesomeCo/Conclusion.yml
extends: occurrence
message: "A level 2 Conclusion heading must exist."
ignorecase: true
scope: raw
level: error
min: 1
max: 1
token: "#{2,} Conclusion"
```

Run Vale against awk.md, and the first error you see is about conclusions:

```
$ vale awk.md
 awk.md
 1:1     error       A level 2 Conclusion heading      AwesomeCo.Conclusion
                     must exist.
 ...
```

This particular rule looked at the whole document for a specific pattern, so Vale wasn't able to tell you what line number and column where the violation occurred, so it'll be at the top of the list.

You can get the position information when you use raw scopes. For example, you can create a rule that looks at the text of links for the words "click here" or "here" and flags them. Since Vale only looks at prose, it won't be able to find hyperlinks, so you can use scope: raw with existence checks and a regular expression that looks for Markdown links.

Add the styles/AwesomeCo/LinkHere.yml file to your project:

```
build_your_style/rules/styles/AwesomeCo/LinkHere.yml
extends: existence
message: "Don't use 'here' or 'click here' as the content of a link."
ignorecase: true
scope: raw
level: error
raw:
  - '\[(click )?here\]\(.+\)'
```

Run Vale against sed.md, and you'll find that the last line of the file has a link with that wording:

```
$ vale sed.md
 sed.md
 ...
 85:181  error    Don't use 'here' or 'click     AwesomeCo.LinkHere
                  here' as the content of a
                  link.
```

When you use the built-in checks with the raw scope, you can't selectively turn those rules off within your document. That's because Vale isn't parsing the document and looking for those comments.

You can exert even more control over how Vale processes things by creating rules that include logic. You'll explore that in Writing Custom Rules with Scripts, on page 58.

Next, you'll look at ensuring consistent capitalization.

Enforcing Capitalization

Google, Microsoft, and many other style guides require headings to use sentence-casing, and both the Microsoft and Google style packages you used in the last chapter include a rule to enforce it:

```
use_existing_styles/multiple/styles/Microsoft/Headings.yml
extends: capitalization
message: "'%s' should use sentence-style capitalization."
link: https://docs.microsoft.com/en-us/style-guide/capitalization
level: suggestion
scope: heading
match: $sentence
indicators:
  - ':'
exceptions:
  - Azure
  - CLI
  - Code
  - Cosmos
  - Docker
```

This rule uses Vale's capitalize check, which has a few options unique to this check. The match option specifies how capitalization works. Vale provides the $sentence, $title, $lower, and $upper variables as shortcuts, but you can also specify a regular expression to enforce casing.

The indicators field lets you specify a list of suffixes that, when found, tell Vale to ignore the next token. In this particular case, this tells Vale to ignore words immediately following a colon in a title.

Add a rule to the AwesomeCo style that enforces title casing for headings. Add styles/AwesomeCo/HeadingTitleCase.yml with the following code:

```
build_your_style/rules/styles/AwesomeCo/HeadingTitleCase.yml
extends: capitalization
message: "'%s' should use title-style capitalization: '%s'"
level: warning
scope:
  - heading
match: $title
style: Chicago
exceptions:
  - macOS
```

When you use $title for the match value, you can specify either Associated Press (AP) or Chicago Manual of Style (CMOS) rules for your titles. Vale defaults to the AP style. When it comes to title casing, both style guides are similar, with the Chicago style letting you lowercase words like "to," "from," "with," and "as" (when it's not used as a conjunction). Pick the title casing style that aligns with your overall style; if your organization does everything with AP style, go that direction. If you're on your own, decide what you like best. If it helps, CMOS almost always uses the Oxford, or serial comma, whereas AP style only requires it when you need to avoid confusion. So if you're pro Oxford comma, your choice is clear.

Once again, you define a message that includes two %s values, so Vale can show found and what it should be according to the rule. Since not every author may be familiar with title casing rules, this output is incredibly helpful. Run Vale against awk.md and test your new rule:

```
$ vale --filter='.Name == "AwesomeCo.HeadingTitleCase"' awk.md

awk.md
 30:4  warning  'Printing specific fields'     AwesomeCo.HeadingTitleCase
                should use title-style
                capitalization: 'Printing
                Specific Fields'
 52:4  warning  'doing Calculations'           AwesomeCo.HeadingTitleCase
                should use title-style
                capitalization: 'Doing
                Calculations'
 70:4  warning  'Wrapping up' should use       AwesomeCo.HeadingTitleCase
                title-style capitalization:
                'Wrapping Up'

✖ 0 errors, 3 warnings and 0 suggestions in 1 file.
```

You have three title issues to fix, but you know exactly how to change them.

The rule you created doesn't handle the document's title because it's in the document's YAML front matter, but you'll fix that next.

Linting Front Matter

The documents in your project have front matter that looks like this:

```
---
title: Data Processing using AWK
summary: Process text data using the awk command
---
```

Vale can process front matter, but the rule you created to enforce heading capitalization doesn't apply to your document title because of the way you scoped the rule. Your current rule only looks at headings:

```
build_your_style/rules/styles/AwesomeCo/HeadingTitleCase.yml
extends: capitalization
message: "'%s' should use title-style capitalization: '%s'"
level: warning
scope:
  - heading
match: $title
style: Chicago
exceptions:
  - macOS
```

To get it to look at the title, adjust your rule to add text.frontmatter.title to the scopes:

```
build_your_style/rules/styles/AwesomeCo/HeadingTitleCase.yml
extends: capitalization
message: "'%s' should use title-style capitalization: '%s'"
level: warning
scope:
  - heading
  - text.frontmatter.title
match: $title
style: Chicago
exceptions:
  - macOS
```

Vale parses front matter into a text.frontmatter scope, and you can access the individual keys within that scope.

Save the file and run Vale again with the same filter:

```
$ vale --filter='.Name == "AwesomeCo.HeadingTitleCase"' awk.md

awk.md
 2:8   warning   'Data Processing using        AwesomeCo.HeadingTitleCase
```

```
 ➤              AWK' should use title-style
 ➤              capitalization: 'Data
 ➤              Processing Using AWK'
  30:4  warning 'Printing specific fields'        AwesomeCo.HeadingTitleCase
                should use title-style
                capitalization: 'Printing
                Specific Fields'
...
```

✖ 0 errors, 4 warnings and 0 suggestions **in** 1 file.

This time, Vale checked the document title as well.

You can test other fields in the front matter the same way. For example, you can ensure that the summary field is less than 160 characters by adding the following rule:

```
build_your_style/rules/styles/AwesomeCo/SummaryLength.yml
extends: occurrence
message: "Try to keep the summary under 160 characters. (%s)."
scope: text.frontmatter.summary
level: suggestion
max: 160
token: .
```

This rule looks like the rule you wrote for checking paragraph lengths, but instead of looking for words, it looks for characters, using . for the token, meaning "any character."

You've built a set of solid rules for your project based on your unique needs. Next, you'll think strategically about how you use existing style rules.

Incorporating and Managing Existing Rules

In the previous chapter, you combined existing packages to lint your prose, and you saw that there was overlap between styles. While you can turn rules off, it may be challenging to maintain your exclusions over time. And those external styles change, so running vale sync may actually change your style rules in a way you didn't expect.

To avoid this, you'll continue to use vale sync to bring in the upstream styles, but then you'll copy the style rules from those packages into your AwesomeCo style. This way, you pick and choose what you want to use without having to automatically adopt another style when it inevitably changes upstream. You get full control over your style rules this way.

If you've followed along with the previous chapter, your Vale configuration includes the following line that brings in the Google and Microsoft style packages, the Readability package, and the alex package:

```
Packages = Google, Microsoft, Readability, alex
```

Run vale sync to ensure your local copies of these packages are up-to-date.

Now you can copy some of those rules into your style.

First, copy the Condescending.yml file from the alex package. This rule contains phrases like "simple" and "obvious" that aren't good choices for educational writing because they are subjective:

```
$ cp styles/alex/Condescending.yml styles/AwesomeCo/
```

Open styles/AwesomeCo/Condescending.yml and look at the rule:

```
build_your_style/rules/styles/AwesomeCo/Condescending.yml
extends: existence
message: Using '%s' may come across as condescending.
link: https://css-tricks.com/words-avoid-educational-writing/
level: error
ignorecase: true
tokens:
  - obvious
  - obviously
  - simple
  - simply
  - easy
  - easily
  - of course
  - clearly
  - everyone knows
```

This is a great list, but it's missing another common word: straightforward. Add that word to the list by adding a new entry to the tokens array:

```
build_your_style/rules/styles/AwesomeCo/Condescending.yml
extends: existence
message: Using '%s' may come across as condescending.
link: https://css-tricks.com/words-avoid-educational-writing/
level: error
ignorecase: true
tokens:
  - obvious
  - obviously
  - simple
  - simply
  - easy
  - easily
```

```
  - of course
  - clearly
  - everyone knows
➤ - straightforward
```

If the upstream file changes in the future, it'll be your responsibility to integrate any changes you want to include, but you're also more in control now because you're using these rules as the foundation for your style guide rather than solely relying on another style guide's maintainers. If things change, run vale sync again and compare the files with a diff tool.

Next, add the Flesch Readability metric by copying it from the Readability style:

```
$ cp styles/Readability/FleschReadingEase.yml styles/AwesomeCo/
```

You previously made a filter in styles/config called readability.exp that looked for rules starting with Readability. This filter won't work now that you've copied the rule into the AwesomeCo style. Modify the filter and change the filter to use the AwesomeCo.FleschReadingEase rule:

```
build_your_style/rules/styles/config/filters/readability.exp
.Name == "AwesomeCo.FleschReadingEase"
```

Run Vale again using the filter:

```
$ vale --filter=readability.exp
```

Finally, copy the rules from the Google and Microsoft packages that check for Oxford comma usage, ellipses, first-person pronouns, and adverbs:

```
$ cp styles/Google/FirstPerson.yml styles/AwesomeCo/
$ cp styles/Google/We.yml styles/AwesomeCo/
$ cp styles/Google/OxfordComma.yml styles/AwesomeCo/
$ cp styles/Microsoft/Adverbs.yml styles/AwesomeCo/
```

Oh, and copy the rule that flags the Latin abbreviations "e.g." and "i.e." since nobody remembers how to use those correctly:

```
$ cp styles/Google/Latin.yml styles/AwesomeCo/
```

Explicitly opting in to rules by adding them to your style is going to be more manageable than including everything and then adding many exclusions to your .vale.ini file or juggling different filters. Look through other rules in the Google or Microsoft style packages and copy the ones you want to use in your style.

Next, you'll create a custom vocabulary so Vale will stop flagging certain words and phrases.

Enforcing Word Usage

Style guides usually have a word usage guide that explains how to refer to product names or other jargon. This ensures that various content production groups use the same vocabulary.

To implement a word list in Vale, you'll use two approaches. First, you'll define acceptable terms that Vale's spellchecker shouldn't flag. Then, you'll create a terms list using Vale's substitution check to help people use the right words when they choose the wrong ones.

First, add Vale to your BasedOnStyles list for the Markdown file types so it starts checking for spelling mistakes again:

build_your_style/vocabulary/.vale.ini
```
[*.md]
BasedOnStyles = AwesomeCo, Vale
AwesomeCo.ParagraphLength = NO
```

This adds spelling errors to the other errors Vale is already flagging, so make a filter that only looks for spelling violations. Create the file styles/config/filters/spelling.exp and add the following line that only looks for rules related to spelling:

build_your_style/vocabulary/styles/config/filters/spelling.exp
```
.Name in ["Vale.Spelling", "Vale.Terms", "Vale.Avoid"]
```

Vale.Spelling looks for misspelled words using Vale's built-in dictionary. The Vale.Terms and Vale.Avoid rules look for words you explicitly allow and reject.

Run Vale against the grep.md tutorial and use this filter to see which words Vale thinks you misspelled:

```
$ vale --filter=spelling.exp grep.md

 grep.md
 16:173   error   Did you really mean           Vale.Spelling
                  'monorepos'?
 36:4     error   Did you really mean 'Nginx'?  Vale.Spelling
 38:1     error   Did you really mean 'nginx'?  Vale.Spelling
 60:119   error   Did you really mean           Vale.Spelling
                  'Monorepos'?
 68:5     error   Did you really mean 'repos'?  Vale.Spelling

✖ 5 errors, 0 warnings and 0 suggestions in 1 file.
```

Vale took issue with the words monorepos, repos, and Nginx. You'll address these with a vocabulary.

> **Joe asks:**
> ## Can I Spell-Check Other Languages?
>
> This book focuses on using Vale with American English spelling, because that's Vale's default behavior. But you can create a rule using another dictionary.
>
> You can use Vale's spelling check[a] and Hunspell-compatible dictionaries like the ones that LibreOffice makes available.[b] You can download the .dic and .aff file for the dictionary you want, place them in the styles/config/dictionaries directory, and use them in a rule like this:
>
> ```
> extends: spelling
> message: "'%s' está mal escrita."
> dictionaries:
>   - en_MX
> ```
>
> This covers basic spell-checking, and you can still use Vale's other rules to cover other cases, since those use regular expressions.
>
> _____
>
> a. https://vale.sh/docs/checks/spelling
> b. https://github.com/LibreOffice/dictionaries

Creating a Vocabulary

You used Vale's substitution rules to enforce word usage, as you did with "utilize" and friends. You'll find that many style guides use this approach to provide suggestions for alternatives. You can separate word usage from your style using Vale's Vocabulary feature. This has some advantages.

First, you can keep the word usage separate from the rest of the style enforcement rules. Second, keeping the vocabulary separate from the style rules means you can share that vocabulary with multiple styles. A company with styles for tutorials may have a different style for blog posts or product releases, but they may need a common word list for both.

The most important benefit is that vocabularies can override some style rules. For example, when you specify certain words in your vocabulary, but they appear in a substitution rule, the vocabulary entry overrides it.

Create a custom vocabulary for AwesomeCo that you can use to allow or reject words in your content. Vocabularies live outside of the style rules and go in the styles/config directory. Create the directory styles/config/vocabularies/AwesomeCo to hold your AwesomeCo vocabulary:

```
$ mkdir -p styles/config/vocabularies/AwesomeCo
```

Now open vale.ini and add this vocabulary to the top of the file using the Vocab key:

```
build_your_style/vocabulary/.vale.ini
StylesPath = styles
MinAlertLevel = suggestion
Packages = Google, Microsoft, Readability, alex
Vocab = AwesomeCo
```

Vale flagged monorepos as a misspelled word in the grep.md file, but this is an acceptable term to use when talking about a repository that contains multiple projects, so create a new text file called accept.txt in the styles/config/vocabularies/AwesomeCo directory with the following line:

```
build_your_style/vocabulary/styles/config/vocabularies/AwesomeCo/accept.txt
[Mm]onorepos?
```

Entries in accept.txt are case-sensitive. You can use regular expressions in the file, so instead of adding all possible variations, add a regular expression that covers the cases you need. In this case, the expression looks for Monorepo, Monorepos, monorepo, and monorepos.

Run Vale against grep.md again and monorepos no longer raises an error:

```
$ vale --filter=spelling.exp grep.md

grep.md
36:4  error  Did you really mean 'Nginx'?  Vale.Spelling
38:1  error  Did you really mean 'nginx'?  Vale.Spelling
68:5  error  Did you really mean 'repos'?  Vale.Spelling

✖ 3 errors, 0 warnings and 0 suggestions in 1 file.
```

Next, you'll ensure consistency with product names.

Enforcing Proper Names

The nginx (pronounced "engine x") is a popular web server and reverse proxy. It's a proper name that you should capitalize consistently throughout the content. The official documentation[2] uses nginx, all lowercase, to refer to the server itself. To keep things consistent in these tutorials, add nginx to the accept.txt file:

```
build_your_style/vocabulary/styles/config/vocabularies/AwesomeCo/accept.txt
[Mm]onorepos?
nginx
```

---

2.   https://nginx.org/en/docs/

When you add entries to accept.txt, they're added to a substitution rule called Vale.Terms that ensures occurrences Vale finds match what you place in the accept.txt file.

Run Vale against grep.md again, and this time you don't see a spelling error for Nginx; you see a Vale.Terms error instead:

```
$ vale --filter=spelling.exp grep.md
```

```
grep.md
36:4  error  Use 'nginx' instead of       Vale.Terms
              'Nginx'.
68:5  error  Did you really mean 'repos'?  Vale.Spelling
```

This time, Vale tells you to use the lowercase nginx spelling because Vale now knows this is a term that it must enforce.

Rejecting Words

Sometimes it's not enough to flag a word as misspelled. Sometimes you want to outright ban a word. For example, you may want to ban the word "repo."

Add the styles/config/vocabularies/AwesomeCo/reject.txt file and an entry that flags both repo and repos:

build_your_style/vocabulary/styles/config/vocabularies/AwesomeCo/reject.txt
```
[Rr]epos?
```

When you add an entry to the reject.txt file, it gets added to an existence check called Vale.Avoid.

This time, when you run Vale against grep.md, you see two errors for repos:

```
$ vale --filter=spelling.exp grep.md
```

```
grep.md
36:4  error  Use 'nginx' instead of       Vale.Terms
              'Nginx'.
68:5  error  Did you really mean 'repos'?  Vale.Spelling
68:5  error  Avoid using 'repos'.         Vale.Avoid
```

Not only does Vale flag it as a spelling error, but it also tells you explicitly that this is a word you need to avoid.

While you can use a substitution check for these kinds of things, the substitution might not be clear. The word "repo" is short for "repository," but it's also short for "repossess." While it's unlikely a technical blog would need to care about that distinction, AwesomeCo has a diverse portfolio of customer use cases, and you can't predict where you might use this vocabulary in the future. That said, try adding it to your badwords.yml file in your style to provide the correction:

```
build_your_style/vocabulary/styles/AwesomeCo/BadWords.yml
extends: substitution
message: "Use '%s' instead of '%s'."
ignorecase: true
level: error
action:
  name: replace
swap:
  - utilize: use
  - utilized: used
  - utilizes: uses
  - utilizing: using
➤ - repo: repository
➤ - repos: repositories
```

Run Vale against grep.md, but this time don't use your filter, since that only shows the spelling errors and won't show your BadWords errors:

```
$ vale grep.md

...

68:5    error        Avoid using 'repos'.           Vale.Avoid
68:5    error        Use 'repositories' instead of  AwesomeCo.BadWords
                     'repos'.
68:5    error        Did you really mean 'repos'?    Vale.Spelling

...
```

This change results in three errors for the same word, but it does get the point across that this word needs changing.

In cases like this, when you have a word you don't want people using at all, add it to the reject list in a vocabulary and move on. You can add a substitution rule in your style for how you want writers to handle it if that's important to you, but there's a reason vocabularies are separate from style rules; vocabularies can override a style's rules.

Overriding Styles with Accepted Terms

Vocabularies are intentionally separate from styles by design. You may share a style with multiple writing projects, and those projects may have their own unique jargon that wouldn't make sense to include globally.

The word list from the Google style guide has a lot of good terms you can build on, but there are a few that might be good to override.

Copy the word list to your project:

```
$ cp styles/Google/WordList.yml styles/AwesomeCo/
```

Run Vale with an inline filter against grep.md that only runs the word list you copied. Remember that it's now called AwesomeCo.WordList since you copied it into the AwesomeCo style directory:

```
$ vale --filter='.Name == "AwesomeCo.WordList"' grep.md

 grep.md
 24:18  warning  Use 'regular expression' instead of 'regex'.  AwesomeCo.WordList
 52:16  warning  Use 'app' instead of 'application'.           AwesomeCo.WordList

✖ 0 errors, 2 warnings and 0 suggestions in 1 file.
```

This flags both regex and application. You could argue that writers should spell out regex, but application is a perfectly fine word to use. Plus, it's not just a "software" application. It could be an employment application or any other kind of usage.

Mark both words as acceptable terms in your vocabulary by adding them to your accept.txt file:

build_your_style/vocabulary/styles/config/vocabularies/AwesomeCo/accept.txt
```
[Mm]onorepos?
nginx
regex
application
```

Run Vale again, filtering results on the word list, and this time Vale won't flag those terms:

```
$ vale --filter='.Name == "AwesomeCo.WordList"' grep.md

✓ 0 errors, 0 warnings and 0 suggestions in 1 file.
```

When you add terms to accept.txt, Vale automatically incorporates them into exceptions, which prevents those checks from firing. For example, you added a rule that checked if an author defined acronyms and abbreviations before using them:

build_your_style/vocabulary/styles/AwesomeCo/FirstUse.yml
```
extends: conditional
message: "Define acronyms and abbreviations on first use. ('%s')"
ignorecase: false
level: suggestion
first: '\b([A-Z]{3,5}s?)\b'
second: '\(([A-Z]{3,5}s?)\)'
exceptions:
  - HTML
  - JSON
  - URL
  - ZIP
  - YAML
```

This rule included exceptions for HTML and JSON. The awk.md file contains the acronym CSV, which Vale flags:

```
$ vale --filter='.Name == "AwesomeCo.FirstUse"' awk.md
```

```
awk.md
 2:30    suggestion   Define acronyms and              AwesomeCo.FirstUse
                      abbreviations on first use.
                      ('AWK')
 6:1     suggestion   Define acronyms and              AwesomeCo.FirstUse
                      abbreviations on first use.
                      ('AWK')
 6:229   suggestion   Define acronyms and              AwesomeCo.FirstUse
                      abbreviations on first use.
                      ('CSV')
38:21    suggestion   Define acronyms and              AwesomeCo.FirstUse
                      abbreviations on first use.
                      ('CSV')
72:1     suggestion   Define acronyms and              AwesomeCo.FirstUse
                      abbreviations on first use.
                      ('AWK')
```

```
✓ 0 errors, 0 warnings and 5 suggestions in 1 file.
```

Instead of adding CSV to the list of exceptions, add it to your accept.txt file:

```
build_your_style/vocabulary/styles/config/vocabularies/AwesomeCo/accept.txt
[Mm]onorepos?
nginx
regex
application
CSV
```

Check the file again, and Vale no longer flags the CSV acronym. It still flags AWK, though:

```
awk.md
 2:30    suggestion   Define acronyms and              AwesomeCo.FirstUse
                      abbreviations on first use.
                      ('AWK')
 6:1     suggestion   Define acronyms and              AwesomeCo.FirstUse
                      abbreviations on first use.
                      ('AWK')
72:1     suggestion   Define acronyms and              AwesomeCo.FirstUse
                      abbreviations on first use.
                      ('AWK')
```

```
✓ 0 errors, 0 warnings and 3 suggestions in 1 file.
```

Vale flags AWK because it meets the criteria of the regular expression in the FirstUse.yml style you created. If you look at the man page for the command, you'll see it's capitalized Awk at the beginning of sentences but in all lowercase everywhere else. In online tutorials, it's presented in a variety of ways.

As a reminder, when there's ambiguity, make a call for consistency in your own documentation. Add AWK to your accept.txt list if you want it capitalized everywhere. If you always want lowercase awk, add it to accept.txt in lowercase instead. If you're fine with both Awk and awk, add [Aa]wk. Finally, remember that Vale won't check things in code font, so you could always refer to awk in a code font consistently and create a scope: raw Vale rule to enforce that.

So far, you've enforced style using regular expressions to look at phrases. Now you'll look at how to catch more complex things that patterns can't reliably catch.

Enforcing Grammar with Natural Language

Most of Vale's rules focus on style, but Vale's sequence check uses part-of-speech tagging to analyze grammar. This means you can create rules that catch issues based on word order, context, or syntax, like passive voice, modal verbs, or adjective-adverb misuse.

To do this, Vale uses the prose[3] package, an English-only Natural Language Processor (NLP), created by the Vale maintainer.

Use the sequence check to create a rule that looks for modals. Modals are verbs that modify the meaning of the main verb to express possibility, necessity, ability, permission, or obligation. For example, "should" and "will" are examples of modals. In docs or educational writing, the present tense is clearer and more direct.

A sequence rule in Vale matches a series of words that follow a specific grammatical pattern. You define "tokens" that define each word's role in the pattern. Create a new file in the styles/AwesomeCo folder called Modals.yml and add this rule that looks for "will" and "should," followed by any verb:

```
build_your_style/vocabulary/styles/AwesomeCo/Modals.yml
extends: sequence
message: "Use present tense instead of '%[1]s %[2]s.'"
level: warning
tokens:
  - tag: "MD"
    pattern: "(will|should)"
  - tag: "VB"
    pattern: ".+"
```

---

3. https://github.com/jdkato/prose

Vale's sequence rules match patterns across sequences of words, rather than a single word, so each token defines the expected part of speech and content for each step in the grammatical structure you want to catch. This rule contains two tokens, each with a tag and pattern.

In the first token, the tag: "MD" matches any modal verb, like "will," "should," "can," "may," and others. The pattern: "(will|should)" narrows the match to "will" or "should." The pipe symbol acts as an "or."

For the second token, the tag: "VB" matches a base-form verb, like "run," "restart," or "configure." The pattern ".+" is a wildcard that matches any word, but since this pattern follows the base verb tag, this pattern matches any base verb.

The message in the rule lets you print the specific tokens in the match. %[1] is the first token's match, and %[2]s is the second.

Run your new rule on sed.md using a filter to test it in isolation:

```
$ vale --filter='.Name == "AwesomeCo.Modals"' sed.md
 sed.md
 6:73    warning   Use present tense instead of      AwesomeCo.Modals
                   'will show.'
 53:13   warning   Use present tense instead of      AwesomeCo.Modals
                   'will change.'

✖ 0 errors, 2 warnings and 0 suggestions in 1 file.
```

This rule found two occurrences. Using a sequence check for this is better than an expects check because you're not just looking for the word "should" on its own; you're looking for it when it precedes a verb.

You can use this approach to scan for more complex situations, such as finding passive voice in your content. Passive voice phrases typically follow this pattern:

```
[form of "be"] + [past participle]
```

In other words, you can spot passive voice by looking for a "be" verb, like "is," "are," "was," "were," "be," "been," or "being," followed by a past participle verb.

For example, "This book was written by me."

Passive voice isn't grammatically wrong, but using active voice instead can make writing clearer. When the subject performs the action, you want the active voice. You'd use passive voice when the subject receives the verb's action, or when the subject doesn't matter.

Passive voice: "This book was written by me." Active voice: "I wrote the book."

As a general rule, technical writers prefer active voice and encourage it in their style guides. Rather than creating a long list of word and phrase combinations to avoid, write a sequence rule. Create a new file called passive.yml in the AwesomeCo style and add this rule to detect passive voice:

```
build_your_style/vocabulary/styles/AwesomeCo/Passive.yml
extends: sequence
message: "Avoid passive voice when possible: '%[1]s %[2]s'"
level: warning
tokens:
  - tag: "VB|VBP|VBZ|VBD|VBG|VBN"
    pattern: "(be|been|being|is|am|are|was|were)"
  - tag: "VBN"
    pattern: ".+"
```

The first token captures the first part of a passive voice phrase: the helping verb that supports a past participle. This rule uses VB again to look for the base form of verbs, along with some new tags:

- VBP matches present tense plural or first or second-person verbs, like "are" or "have."
- VBZ matches third-person singular present tense verbs, like "is" or "has."
- VBD matches past tense verbs, like "was" or "had."
- VBG matches present participle verbs, also known as gerunds. These are verbs ending in "-ing."
- VBN matches past participle verbs, like "written" or "been."

The pattern for the first token then limits the match to specific forms of "be" that start a passive construction.

The second token matches the main verb in passive voice, following the auxiliary verb. tag: VBN matches past participle verbs again, and the pattern matches any verb.

Scan the sed.md file with Vale. Use a filter to show the results of your new rule:

```
$ vale --filter='.Name == "AwesomeCo.Passive"' sed.md

sed.md
6:194  warning  Avoid passive voice when        AwesomeCo.Passive
                possible: 'is used'

✖ 0 errors, 1 warning and 0 suggestions in 1 file.
```

Using a combination of tags and patterns, you can construct all kinds of grammar and style rules without writing large collections of regular expressions.

Your Turn

Complete the following challenges to deepen your understanding of how to build custom style rules:

1. Add other words to your BadWords.yml that you want to avoid in your writing. Test them out with a new Markdown document.
2. You have a rule for the length of your headings. Create a similar rule that looks at sentence length to ensure it isn't more than 25 words.
3. Extend your vocabulary to make sure you consistently use grep, sed, and awk in lower case across the tutorials.
4. Explore creating "unit tests" for your custom rules. Write Markdown documents that intentionally contain style violations that you can run Vale against to ensure your rules fire as expected. You can use a library like shellspec[4] or write your own scripts that parse the output from Vale. Remember that you can tell Vale to render its output as JSON, too.

Wrapping Up

You can now write several kinds of Vale rules from scratch. You can look for individual words, define a custom vocabulary, write rules to enforce grammar, and script the stuff you can't do any other way. And you know you don't have to write it all from scratch. The most effective way you can create your own style is to build as much of it as you can from existing styles and then add your own customizations on top. Use a vocabulary to create exceptions to terms in substitution or existence rules when you can, and don't hesitate to edit the styles you copy into your project. You can always redownload the originals later if you change your mind.

Next, you'll look at integrating Vale with your editor, supporting other formats, and using Vale with other tools.

---

4. https://github.com/shellspec/shellspec

CHAPTER 4

Integrating Vale into Your Workflow

Now that you have Vale enforcing your style, you'll want to make Vale an integral part of your workflow. You'll want it to work seamlessly as you write your docs, checking things as you write. You'll want it as part of your documentation publication process when you make pull requests on GitHub. You'll certainly want to share the style you created across your various writing projects, and you'll want to handle some custom formats in those rare cases where Vale can't interpret your code.

Start by wiring Vale up to your text editor so you can lint prose while you create it.

Seeing Errors as You Work

If you're working on a software project, there's a good chance the project has some sort of style enforcement. JavaScript and TypeScript projects use ESLint,[1] Ruby projects use Rubocop,[2] and Python has several linting options available. Those linting tools integrate with editors so you can see issues as they happen rather than waiting until it's time to push your code to staging.

You can work Vale into your editor too.

Using Vale with Visual Studio Code

To use Vale with Visual Studio Code, you install and configure the vale-vscode extension:[3] press Ctrl+P to bring up the Command Palette and enter the following command to install the extension:

```
ext install ChrisChinchilla.vale-vscode
```

---

1.   https://eslint.org/
2.   https://rubocop.org/
3.   https://marketplace.visualstudio.com/items?itemName=ChrisChinchilla.vale-vscode

Once installed, you'll see errors as you work. The Vale extension automatically locates the .vale.ini file relative to the file you're opening. Open awk.md and you'll find errors underlined in the file. Issues like your Flesch Reading Ease rule and your conclusion rule show up as an issue on the first line. Hover over the error marker to see the errors:

If your Vale rules include action commands, they'll appear as quick-fix actions:

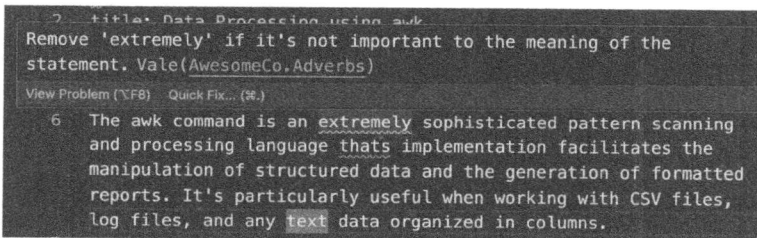

Finally, Vale rules show up under the Problems tab:

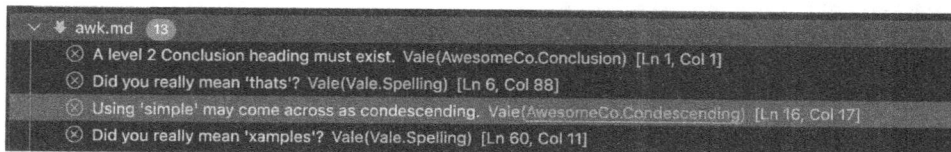

Now you can catch errors as you go, without having to switch context.

Vale and the Language Server Protocol

Language Server Protocol (LSP) is an open standard that lets any code editor talk to a language-specific server so you get autocompletion, real-time error checking, and other benefits that help you work faster. Vale has LSP support, which is how the Visual Studio Code extension works. Other editors that support LSP can use Vale as well. For example, the Zed[4] editor has a Vale LSP plugin[5] you can install to get inline errors.

Any editor that supports LSP can integrate with Vale this way.

---

4.   https://zed.dev/
5.   https://github.com/koozz/zed-vale

Using Vale with Vim and Neovim

You can get in-line feedback in Vim and Neovim with either Ale[6] or CoC.[7] Both options work well and can display error messages inline.

If you use Ale, and Vale is on your path, things "just work." Ale executes Vale by using a temporary file, though, so if you have a .vale.ini file with a pattern that uses directories in the file patterns, you won't see errors because Vale won't match the temporary file with the specific pattern in its configuration. Avoid this by using patterns like you've used in this book, where you use directory-specific configuration files with patterns that only look at extensions.

For CoC, you'll need to do a few extra steps. First, install the coc-vale plugin. In Vim's Command Mode, run this command:

```
:CoCInstall coc-vale
```

Then, if you want errors to show inline, you need to make sure that you've enabled virtual text and diagnostic messages. Open CoC's configuration file with the following command in Vim's Command Mode:

```
:CoCConfig
```

Then add the following lines to the file:

```
{
  "diagnostic.virtualText": true,
  "diagnostic.enable": true,
  "diagnostic.virtualTextCurrentLineOnly": false,
  "vale.valeCLI.minAlertLevel": "suggestion"
}
```

With this in place, errors show inline:

---

6.   https://github.com/dense-analysis/ale
7.   https://github.com/neoclide/coc.nvim

You can use CoCList diagnostics to see all the errors:

```
>>  1 |-- A level 2 Conclusion heading must exist. Try to keep the Flesch reading ease score ...
    2 A level 2 Conclusion heading must exist. (Vale AwesomeCo.Conclusion)
    3
    4 Try to keep the Flesch reading ease score (52.45) above 70. (Vale AwesomeCo.
    5 FleschReadingEase)
>>  6 The awk command is an extremely sophisticated pattern scanning and processing language th
      ats implementation facilitates the manipulation of structured data and the generation of
      formatted reports. It's particularly useful when working with CSV files, log files, and a
      ny text data organized in columns.  Did you really mean 'thats'? Remove 'extremely' if i...
    7
    8 ## Basic Structure
    9
   10 The basic structure is:
   11
```

As with Visual Studio Code, you can see errors as you create them, which speeds up the review process considerably.

Getting Vale to show errors while you work is a great improvement over running the CLI tool whenever you make changes.

So far, you've used Vale on your prose. But sometimes you have prose inside of a codebase.

Joe asks:
Is Vale Still Useful When We Have LLMs and Coding Assistants?

Large Language Models (LLMs) are accelerating how people produce code and content, so it's only natural to wonder if Vale is still relevant. While LLMs continue to evolve, they have limitations that still make Vale an important part of your content-creation toolbox.

You can use LLMs to edit and fix your content, but you may find that they change more than you asked for, or even rewrite your content in subtle ways you didn't expect. You'll still have to prompt your LLM with all of your style rules anyway. Even with the strictest instructions, they still struggle to follow every rule. Watch what they change and ensure it's what you want.

In addition, LLMs are nondeterministic, which means you get different results each time you run a query. When you're producing content, you want consistency.

You'll get the best approach if you use Vale alongside your LLMs. Take drafts you create and run them through your style rules. If you use a coding assistant powered by an LLM, make sure it knows how to run Vale like it would run the rest of your test suite.

LLMs are constantly improving, so periodically explore how Vale fits into your workflow as things evolve.

Linting Comments in Source Code

Engineers often write detailed comments in source code and then use tooling to generate API or SDK documentation from that code. Sometimes these end up as official pieces of documentation. You can use Vale to ensure that this documentation has the same level of consistency as your other content. You've only looked at linting prose in content files, but Vale can scan comments in several programming languages, including Go, Ruby, Rust, Java, C, Java-Script, and even CSS.

Test this out by creating a small web server in the Go programming language. You won't need to run this program to test out Vale's ability to check the comments for errors, but it is a functioning server you can build and run if you have Go installed. That's outside the scope of the book.

Create a new file called server.go that contains the following code:

```
integrating/codelint/server.go
// A basic web server
package main

// Import the pakcages
import (
        "log"
        "net/http"
)

/* Starts server and serves files from the the current dir.
 * listens on 0.0.0.0 port 1337.
 */
func main() {
        port := "1337"
        // Utilize http.FileServer to serve files from the current dir (.).
        fs := http.FileServer(http.Dir("."))
        http.Handle("/", fs) // serv all files

        log.Printf("Serving on http://0.0.0.0:%s...", port)
        err := http.ListenAndServe("0.0.0.0:"+port, nil)
        if err != nil {
                log.Fatal(err)
        }
}
```

This small program acts as a static file server, serving any files in the local directory. Notice that the comments in the code have some issues, though.

Tell Vale to look at this file for issues that make sense for code. For example, since code won't have conclusion headings, turn that rule off. Open .vale.ini and add a new section for files with the .go extension:

```
integrating/codelint/.vale.ini
StylesPath = styles
MinAlertLevel = suggestion
Vocab = AwesomeCo
Packages = Google, Microsoft, Readability, alex

[*.md]
BasedOnStyles = AwesomeCo, Vale
AwesomeCo.ParagraphLength = NO

[*.go]
BasedOnStyles = AwesomeCo, Vale
AwesomeCo.Conclusion = NO
```

Now run Vale against server.go:

```
$ vale server.go
 server.go
 4:15    error   Did you really mean 'pakcages'?   Vale.Spelling
 10:40   error   'the' is repeated!                Vale.Repetition
 15:5    error   Use 'use' instead of 'Utilize'    AwesomeCo.BadWords
 17:26   error   Did you really mean 'serv'?       Vale.Spelling

✖ 4 errors, 0 warnings and 0 suggestions in 1 file.
```

Vale caught the bad word, some spelling errors, and the repeated "the" in the sentence.

Using Vale to scan your docs and your code helps you build a consistent voice across all of your content, and since you can use Vale with code editors, engineers can have Vale lint their comments as they write them.

Your style covers a lot of common cases, but you may run into situations where Vale's built-in rules aren't enough. If you encounter a situation where Vale doesn't support what you're trying to do, you can parse the raw file yourself.

Writing Custom Rules with Scripts

When you're limited by a basic Vale expression and need more flexibility, you can use the script check, which lets you write your own processing logic against the raw content of the document. You'll need experience with scripting languages, but if you can code, this is a way you can extend Vale to handle edge cases in your content.

You write scripts in the Tengo[8] language and either include them directly in the YAML file or in their own script in your styles/config/ directory. Tengo is a small scripting language that's secure and offers many text-processing functions. Vale can run Tengo scripts directly, so you don't have extra dependencies to install.

---

8.   https://github.com/d5/tengo

The following code listing shows a basic Tengo script that counts the characters in a string and prints the string and characters to the screen:

```
// Imports
fmt := import("fmt")
text := import("text")

// define variables
name := "Barney"
chars := len(name)

// build strings
result := "Hello, " + name + ". Your name has " + chars + " characters."

fmt.println(result)
```

Visit the Tengo Playground[9] and enter this script to test it yourself.

You can use Vale with Tengo to solve a common issue technical writers face when working with code snippets. In technical documentation, it's a good idea to keep your code block lines under a certain length to improve readability. It's especially important if you're going to present your content in print or in a PDF format where readers can't scroll horizontally. You won't be able to use Vale to scan code blocks because Vale ignores them. But you can use the raw scope and a script to locate the code blocks and then count the characters in those blocks.

Your script will process your content file line by line. You'll use a regular expression to look for the three backticks that represent the start of a Markdown code block and set a variable to denote you're inside of the code fence. When you're in the code fence, you'll get the length of the current line. If the line is too long, you'll return the starting and ending character positions to Vale by creating an array called matches containing all the matches you find.

Create a rule called CodeLineLength.yml in the styles/AwesomeCo folder. Add the following code to the file:

```
integrating/tengo/styles/AwesomeCo/CodeLineLength.yml
extends: script
message: "Try to keep code line length under 80 characters (%s)."
scope: raw
level: warning
script: |
  fmt := import("fmt")
  text := import("text")

  matches := []
  cursor := 0
```

---

9. https://tengolang.com/

```
in_code_block := false
max_line_length := 80

for line in text.split(scope, "\n") {
  if text.re_match("^```.*$", line) {
    in_code_block = !in_code_block
  } else if in_code_block {
    if len(line) > max_line_length {
      matches = append(matches, {
        begin: cursor,
        end: cursor + len(line),
        text: line
      })
    }
  }

  // Update cursor position after processing each line
  cursor += len(line) + 1  // +1 for the newline character
}
```

The rule itself specifies that it's a script and that it's looking at the raw text, so Vale won't do any preprocessing. The Tengo script is under the script key.

In the script itself, you include the text library and declare variables for the array of matches you'll return to Vale, the length you want to look for, and a variable called cursor that keeps track of your current character position in the document. As you read through each line, you'll add the length of that line to the cursor.

The scope variable contains the text from the scope. Since this rule's scope is raw, scope contains the entire file.

You split the entire file into lines and process each line, matching it against the regular expression that looks for the code block. If you find a match, you set the in_code_block variable.

On the next line, you check if you're in a code block. If you are, you check the length of the current line. If the length is too long, add an entry to matches with the begin and end positions of the match:

`integrating/tengo/styles/AwesomeCo/CodeLineLength.yml`
```
if len(line) > max_line_length {
  matches = append(matches, {
    begin: cursor,
    end: cursor + len(line),
    text: line
  })
```

Since you're processing the file line by line, you have to use the cursor value in the begin and end positions so that Vale knows exactly what character in the whole file the match starts on and where the match ends. While you're processing the text line by line for convenience, Vale is looking at the file as a single string.

Finally, you move the cursor forward by the current line's length plus 1 so subsequent lines' matches have correct absolute start and end positions.

Now test the new rule. Create a new text file called longline.md with a long code line:

integrating/tengo/longline.md
```
This is a document with a long code line.

```
```
This is a long line of text in of a code fence in Markdown and triggers the warning.
```

Then run Vale with a filter to test the new rule in isolation:

```
$ vale --filter='.Name == "AwesomeCo.CodeLineLength"' longline.md
longline.md
 4:1  warning   Try to keep code line length      AwesomeCo.CodeLineLength
                under 80 characters (This is a
                long line of text in of a code
                fence in Markdown and triggers
                the warning.).

✖ 0 errors, 1 warning and 0 suggestions in 1 file.
```

The long code line gets flagged.

You don't have to embed the script in the YAML file. You can point to an external script file that you place inside of your styles/config/scripts folder. For example, put the script in a file called CodeLineLength.tengo in styles/config/scripts. Then, in your YAML rule, set the script value to the name of the file.

Try it out. Change your YAML file to point to a script called CodeLineLength.tengo:

integrating/tengo-split/styles/AwesomeCo/CodeLineLength.yml
```
extends: script
message: "Try to keep code line length under 80 characters (%s)."
scope: raw
level: warning
script: CodeLineLength.tengo
```

Then make a new scripts folder under your style/config directory:

```
$ mkdir style/config/scripts
```

In that folder, create the file CodeLineLength.tengo and place the content of your Tengo script there.

```
integrating/tengo-split/styles/config/scripts/CodeLineLength.tengo
fmt := import("fmt")
text := import("text")

matches := []
cursor := 0

in_code_block := false
max_line_length := 80

for line in text.split(scope, "\n") {
    if text.re_match("^```.*$", line) {
      in_code_block = !in_code_block
    } else if in_code_block {
      if len(line) > max_line_length {
        matches = append(matches, {
          begin: cursor,
          end: cursor + len(line),
          text: line
        })
      }
    }

    // Update cursor position after processing each line
    cursor += len(line) + 1  // +1 for the newline character
}
```

If you copied this from your YAML file, remove all the leading spaces from the file.

Now run Vale against longline.md again to ensure your rule still works:

```
$ vale --filter='.Name == "AwesomeCo.CodeLineLength"' longline.md

longline.md
 4:1  warning  Try to keep code line length      AwesomeCo.CodeLineLength
                under 80 characters (This is a
                long line of text in of a code
                fence in Markdown and triggers
                the warning.).

✖ 0 errors, 1 warning and 0 suggestions in 1 file.
```

Having the script in a separate file can make things feel more organized, but there aren't significant advantages beyond that unless you start building more sophisticated rules you'd like to test. In most cases, you can test your Tengo scripts directly in the Tengo playground by setting the scope value in the script to a string of test content and by using fmt.println to print out any debug messages you want to display. Choose the approach that makes sense for your projects.

Vale continues to evolve, adding support for more formats, but you may encounter exceptions or strange edge cases.

Working with Custom Formats

Vale understands how to parse Markdown, AsciiDoc, reStructuredText, XML, and many programming languages. Vale does its best to process these files and only lint the prose by identifying text elements. Some content management systems introduce proprietary markup or use formats that Vale struggles to understand.

You can handle these situations in a few ways.

First, you can map one format to another. For example, if you have Markdown files, but you've saved them with the txt extension, you can use the formats section in .vale.ini to map one type to the next.

```
[formats]
txt = md

[*.txt]
; your specific rules
```

If that's not enough, then you can tell Vale to ignore things. The easiest way to do that is to add comments in your documents that turn Vale on and off for certain pieces.

In Markdown and HTML documents, you can use <!-- vale off --> to tell Vale to stop processing some content. Then use <!-- vale on --> to enable it again.

You can also turn specific rules on and off. If you have a passage where your back is against the wall and you must use "utilize" in your content, you can turn the rule off before you use the text and turn it on after:

```
<!-- vale AwesomeCo.BadWords = NO -->
You can also turn specific rules on and off. If you have a passage where your
back is against the wall you and must use "utilize" in your content,
you can turn the rule off before you use the text, and turn it on after:
<!-- vale AwesomeCo.BadWords = YES -->
```

This approach won't work with rules that use the raw scope, and it won't work with your Tengo script rules unless you look for those comments specifically using a pattern like this:

```
integrating/tengo-split/styles/config/scripts/CodeLineLength.tengo
fmt := import("fmt")
text := import("text")

matches := []
cursor := 0
ruleActive := true

in_code_block := false
```

```
  max_line_length := 80
  for line in text.split(scope, "\n") {
    if text.re_find(`<!-- vale AwesomeCo.CodeLineLength = NO -->`, line, -1) {
      ruleActive = false
    }

    if text.re_find(`<!-- vale AwesomeCo.CodeLineLength = YES -->`, line, -1) {
      ruleActive = true
    }
    if ruleActive {
      if text.re_match("^```.*$", line) {
        in_code_block = !in_code_block
      } else if in_code_block {
        if len(line) > max_line_length {
          matches = append(matches, {
            begin: cursor,
            end: cursor + len(line),
            text: line
          })
        }
      }
    }
  }

  // Update cursor position after processing each line
  cursor += len(line) + 1  // +1 for the newline character
}
```

As you read each line, look for the comments and keep track of whether the rule is on or not. Run your parsing logic only if the rule is active.

If adding these exclusions to your document won't scale, you can use the BlockIgnores and TokenIgnores fields in your .vale.ini file and specify regular expressions for sections you want Vale to ignore. These work in Markdown, AsciiDoc, reStructuredText, and Org Mode content, and require you to write regular expressions that capture the entire block or token in the first grouping.

For example, you might be using attributes in your Markdown to add classes and IDs to your headings, like this:

```
## Working with Custom Formats {#sec.custom.format .majorHeading}
```

When you use a rule that enforces title casing, Vale would consider the attributes part of the title and flag this as an error.

To get around this, add a TokenIgnores rule to the Markdown section of your .vale.ini file that looks for classes or IDs inside the curly braces:

```
[*.md]
TokenIgnores = (\{[ .#][^}]+\})
...
```

Notice that the entire expression is within a capture group, which is what the outermost parentheses represent. Whatever this regular expression captures is what Vale ignores.

The `BlockIgnores` feature is for ignoring blocks of text, while `TokenIgnores` is for ignoring inline text. Both take a single regular expression or several expressions separated by commas, with each regular expression using a capture group to tell Vale what to ignore.

These work for basic structures, but they don't work well when parsing structured content.

Working with MDX Files

If you're using documentation platforms, like Docusaurus[10] or Eleventy,[11] or other content systems based on Next.js, you've encountered MDX, a popular file format that lets you mix Markdown and JSX syntax.

MDX is flexible. You can add interactive elements to your content, like buttons, slideshows, diagrams, and other components you've used throughout your site. You can also treat MDX files as components themselves, so you can include one page in another, which is a great way to reuse content.

Here's an example of an MDX file:

```
---
title: Sample MDX file
description: This is a sample.
---

import {Button} from "./buttons";

This is regular Markdown, but it has a call-to-action button:

<Button text="Learn more" />

This is more Markdown text again.

Last updated {new Date().getFullYear()}
```

In this example, you see a JavaScript `import` statement, regular Markdown text, and a React `Button` component. Typically, the framework you use with MDX runs a preprocessor that converts the files to regular HTML files and either writes them to disk or serves them through the framework.

The naive way of getting Vale to work with MDX files is to use the `formats` option and map Markdown files to MDX files:

---

10. https://docusaurus.io/
11. https://www.11ty.dev/

```
[formats]
mdx = md
```

This often isn't enough, so people try using Vale's TokenIgnores and BlockIgnores options, using regular expressions to identify chunks of code to skip. When you do this, you'll find that complex JSX elements and embedded object structures trip Vale up. Vale may lint things it shouldn't or ignore things you want it to parse. React components are essentially XML, and it's not a good idea to try to parse XML or HTML with regular expressions.

To get Vale to play nicely with MDX files, you install an external command-line tool that converts MDX files to a format Vale can use. This tool requires Node.js, but if you're using MDX files, you're already using Node.js as part of your development environment anyway.

Install the tool with npm:

```
$ npm install -g mdx2vast
```

Once you've installed the tool, double-check that you're using mdx in your .vale.ini file instead of md when defining the style rules you'll use:

```
[*.mdx]
BasedOnStyles = AwesomeCo, Vale
AwesomeCo.ParagraphLength = NO
```

Then ensure you aren't mapping mdx files to another format.

The only other change you'll need to make is how you turn Vale rules on and off in your documents instead of using regular HTML comments, like this:

```
<!-- vale AwesomeCo.BadWords = NO -->
```

You need to use MDX comments:

```
{/* vale AwesomeCo.BadWords = NO */}
```

If you've written custom Tengo scripts that look for HTML comments, you'll need to update those as well if they target HTML comments.

Now, when you run Vale, it uses the mdx2vast command to transform the document into an HTML document that Vale can process. This intermediate step increases the total linting time of large sites, but you won't notice it when you're working on an individual file.

Depending on how complex your MDX content gets, you may still run into false positives.

Using Vale with Hugo

Hugo[12] is a static site generator that can use Markdown files. Hugo lets you create shortcodes, which are similar to React components. You insert these shortcodes into your Markdown files using a certain syntax.

Vale might get tripped up when it encounters these shortcodes in your files, so you can tell Vale how to ignore them. Rather than writing your own regular expressions, you can use the vale-hugo package,[13] which already has those written.

Add the following package to your Vale configuration:

```
Packages = Hugo
```

Then run vale sync to download the package.

Vale packages don't have to contain style rules. They can contain a .vale.ini file with settings. This particular configuration file includes the regular expressions you need to parse out Hugo shortcodes.

Parsing Other Formats with Views

Vale's views[14] give you a way to transform structured documents into scopes Vale can understand. Using views, you can zero in on areas of a document you care about, while ignoring the rest.

In Node.js applications, the package.json file includes all the information about a project, including its name, version, and dependencies. It also includes a description that you'll want to check with Vale.

Create a package.json file for a basic web server. You won't implement the server, but you'll want a realistic file you'll encounter in the wild:

```
integrating/views/package.json
{
  "name": "webserver",
  "version": "1.0.0",
  "description": "Simple web server that converts Markdown to HTML files.",
  "keywords": [
    "web",
    "server",
    "Markdown",
    "HTML"
  ],
```

---

12. https://gohugo.io/
13. https://github.com/errata-ai/Hugo
14. https://vale.sh/docs/views

```
  "license": "MIT",
  "author": "Brian P. Hogan",
  "type": "commonjs",
  "main": "app/server.js",
  "scripts": {
    "test": "echo \"Error: no test specified\" && exit 1"
  }
}
```

Now add a section to your .vale.ini file that looks at package.json files and applies the Vale and AwesomeCo styles:

integrating/views/.vale.ini

```
[package.json]
BasedOnStyles = AwesomeCo, Vale
```

Now run Vale against your package.json file. It flags more than the description:

```
$ vale package.json
```

```
package.json
 1:1    error      A level 2 Conclusion heading    AwesomeCo.Conclusion
                   must exist.
 2:12   error      Did you really mean             Vale.Spelling
                   'webserver'?
 4:19   error      Using 'Simple' may come across  AwesomeCo.Condescending
                   as condescending.
11:15   suggestion Define acronyms and             AwesomeCo.FirstUse
                   abbreviations on first use.
                   ('MIT')
13:12   error      Did you really mean             Vale.Spelling
                   'commonjs'?
```

✖ 4 errors, 0 warnings and 1 suggestion in 1 file.

To get around this, create a view that only looks for the description field.

A view consists of a transformation engine and a series of transformation steps. You choose the engine based on what you want to parse. Use Dasel[15] to parse YAML, JSON, and TOML files and tree-sitter[16] to parse code. Then you write queries to get the pieces you want.

Create the views directory in config/styles:

```
$ mkdir styles/config/views
```

---

15. https://github.com/TomWright/dasel
16. https://tree-sitter.github.io/tree-sitter/

Then create the PackageView.yml file with the following contents:

```
integrating/views/styles/config/views/PackageView.yml
engine: dasel
scopes:
  - name: description
    expr: description
    type: md
```

Since you're parsing JSON, you'll use Dasel as the engine. Each step in the View includes a name, an expression, and the content type. Since you're only looking for one field, you only have one step.

Now that you have the view, add a reference to the view in your .vale.ini file within the package.json section, and also turn off the Conclusion and ParagraphLength rules since those don't apply to JSON files:

```
integrating/views/.vale.ini
[package.json]
BasedOnStyles = AwesomeCo, Vale
➤ View = PackageView
➤ AwesomeCo.ParagraphLength = NO
➤ AwesomeCo.Conclusion = NO
```

Now run Vale against your package.json file, and it only returns the error in the description field:

```
$ vale package.json
 package.json
 4:19  error  Using 'Simple' may come across   AwesomeCo.Condescending
              as condescending.

✖ 1 error, 0 warnings and 0 suggestions in 1 file.
```

Explore Vale's documentation for more details on parsing more complex files.

You now have the tools to get Vale to parse custom elements and work around parsing issues.

Vale is all about helping you create consistent, quality content. Another way to ensure consistency is to make sure all of your projects use the same styles.

Sharing Your Style across Multiple Projects

You installed existing packages using vale sync, and as you explored the packages, you discovered that they were collections of style definitions. You can turn your custom style into a package others can use as well.

A Vale package is a Zip file containing a styles folder, a .vale.ini file, or both. You can include your vocabulary, filters, and even your file mappings, which means you can maintain your style in one place but use it on multiple content repositories without having to worry about things getting out of sync.

To package the entire AwesomeCo style you created into a Package, you'll place your .vale.ini file and your styles/config and styles/AwesomeCo folders in a new directory.

First, create a new AwesomeCo directory with a styles folder:

```
$ mkdir -p AwesomeCo/styles
```

Move your .vale.ini file into that new AwesomeCo folder:

```
$ mv .vale.ini AwesomeCo/
```

Then move styles/AwesomeCo and styles/config folder into AwesomeCo/styles:

```
$ mv styles/AwesomeCo AwesomeCo/styles/AwesomeCo
$ mv styles/config AwesomeCo/styles/config
```

The directory structure for your package now looks like this:

```
AwesomeCo/
├── .vale.ini
└── styles
    ├── AwesomeCo
    │   ├── Adverbs.yml
    │   └── other files
    └── config
        ├── filters
        │   ├── readability.exp
        │   └── other files
        ├── scripts
        │   └── CodeLineLength.tengo
        ├── views
        │   └── PackageView.yml
        └── vocabularies
            └── AwesomeCo
                ├── accept.txt
                └── reject.txt

9 directories, 27 files
```

Now open the new AwesomeCo/.vale.ini file and remove the Packages line so your file looks like this:

```
integrating/share/AwesomeCo/.vale.ini
StylesPath = styles
MinAlertLevel = suggestion
Vocab = AwesomeCo

[*.md]
BasedOnStyles = AwesomeCo, Vale
AwesomeCo.ParagraphLength = NO
```

```
[*.go]
BasedOnStyles = AwesomeCo, Vale
AwesomeCo.Conclusion = NO

[package.json]
BasedOnStyles = AwesomeCo, Vale
View = PackageView
AwesomeCo.Conclusion = NO
AwesomeCo.ParagraphLength = NO
```

You could leave those packages in, and Vale would install those packages when it installs your package. But it's best to let people install additional packages on their own.

Notice you're leaving in the Markdown file definition. While you could leave this out too, this demonstrates that you can ship a style with default rule associations.

Now create a Zip file of the AwesomeCo folder called AwesomeCo.zip that includes the base folder and its contents:

```
$ zip -r AwesomeCo.zip AwesomeCo/

  adding: AwesomeCo/ (stored 0%)
  adding: AwesomeCo/styles/ (stored 0%)
  ...
  adding: AwesomeCo/styles/AwesomeCo/Conclusion.yml (deflated 21%)
  adding: AwesomeCo/.vale.ini (deflated 49%)
```

To use this package, you reference the Zip file in the Packages section of your .vale.ini file. Since you moved your .vale.ini into the style, create a new one that includes the package:

```
integrating/share/.vale.ini
StylesPath = styles
MinAlertLevel = suggestion
Packages = AwesomeCo.zip

[*.md]
BasedOnStyles = AwesomeCo, Vale
AwesomeCo.ParagraphLength = NO
```

Since the AwesomeCo style's .vale.ini contains your vocabulary, you can omit the Vocab key as well. And since it contains the rules for Markdown files, you can omit that section as well.

Run vale sync, and the AwesomeCo package gets extracted and copied to your styles directory.

```
Syncing AwesomeCo [1/1] ███████████████████████  100% | 0s
 SUCCESS  Synced 1 package(s) to '/Users/brianhogan/vale/styles'.
```

Your styles directory once again contains an AwesomeCo folder, and the files from the style's config folder, including its filters and vocabularies merged into your styles/config folder. You'll also find a new .vale-config folder containing the 0-AwesomeCo.ini file, which is the .vale.ini file you packaged in the style.

Vale combines all the styles it finds together, and as you can guess, if there are conflicting files or rules, you can have collisions. This is where the loading order matters.

To share this style with other projects, publish the Zip file online. Then modify your .vale.ini file to point to the full URL to the Zip file. When you run vale sync, Vale downloads and extracts the style. Now anyone in the world can use your style.

You've gotten Vale working locally, you've integrated it with your processes, and you've created a reusable style. Now you'll integrate Vale into your build process.

Using Vale with GitHub Actions

Running Vale locally is a great way to catch issues as they happen, but you might not want to rely on everyone running it locally, especially if you run a project that has many external contributors. If you use GitHub, you can use the official Vale GitHub Action[17] to lint files whenever you create a GitHub pull request.

To use it, add a .github/workflows folder to your project:

```
$ mkdir -p .github/workflows
```

Then create .github/workflows/vale.yml and add the following code:

.github/workflows/vale.yml

```
name: Vale
on: [pull_request]

jobs:
  vale:
    name: Run Vale on the book
    runs-on: ubuntu-latest
    steps:
      - uses: actions/checkout@v4
      - uses: errata-ai/vale-action@v2.1.1
        with:
          version: 3.10.0
```

Add the .github/workflows/ directory, commit, and push your code:

---

17. https://github.com/errata-ai/vale-action

```
$ git add .github/workflows
$ git commit -m "add Vale workflow"
$ git push origin main
```

Now, when you make a pull request, you'll get annotations added to the GitHub pull request related to the things Vale found.

This GitHub Action relies on reviewdog[18] to do the actual error processing and reporting. It's only going to report things you altered in the commit. To get it to show you more errors than just what changed, add the filter_mode option:

```
name: Vale
on: [pull_request]

jobs:
  vale:
    name: Run Vale on the book
    runs-on: ubuntu-latest
    steps:
      - uses: actions/checkout@v4
      - uses: errata-ai/vale-action@v2.1.1
        with:
          version: 3.10.0
          filter_mode: nofilter
```

When you set filter_mode to nofilter, it scans every file, even if it wasn't part of the pull request. It adds those annotations to the Files tab on the pull request, below the usual diff view.

If that's too noisy, write your own custom action that runs Vale on your content directly, relying on Vale's return value to stop the build.

The following action downloads and runs Vale on every push to the repository, checking every file:

```
name: Vale
on: [push]
jobs:
  vale:
    name: Run Vale on the whole project
    runs-on: ubuntu-latest
    steps:
      - uses: actions/checkout@v4
      - name: Install Vale
        run: |
          wget https://github.com/errata-ai/vale/releases/download/
v3.13.0/vale_3.13.0_Linux_64-bit.tar.gz
```

---

18. https://github.com/reviewdog/reviewdog

```
        tar -xf vale_3.13.0_Linux_64-bit.tar.gz
        sudo mv vale /usr/local/bin/
    - name: Run Vale
      run: vale .
```

Note that the wget line in this example is broken into two lines for readability. Ensure you combine those lines in your GitHub Action.

If Vale returns an error, the action fails, which is enough to prevent your pull request from moving forward. You can then rely on GitHub's notifications, or you can add more steps to the action to make sure the right people get notified. And you can review the Action's results to see the Vale errors that need correcting. But remember that this approach only breaks the build when Vale rules result in errors, not warnings or suggestions. While you can review all Vale errors in the Action's results, nothing will prevent you from merging the pull request with this approach.

Your Turn

1. Take an existing project containing either Markdown or HTML files and add Vale to it. Use the package you created and see how many things get flagged.
2. Explore other style packages. In the appendix, you'll find a list of publicly-available Vale rules from a variety of projects.
3. Explore Vale's templates[19] feature to gain more control over how Vale outputs its results.

Wrapping Up

You now know how to use Vale to enforce your own style guide. You can use and adapt existing rules, write your own rules from scratch, and incorporate Vale into various scenarios.

Vale continues to evolve, adding new features and improving its capabilities. This book covered the features you need to implement Vale in your project; consult the official documentation on how to use Vale's other features.

Once you fully integrate Vale into your workflows, your content will be more consistent, you'll have fewer arguments during peer-review time, and you'll publish faster.

---

19. https://vale.sh/docs/templates

Resources Referenced

The following third-party repositories provide valuable insight into how Vale rules work and how different organizations use Vale to enforce their style:

- Grafana Writers Toolkit[1]
- RedHat Vale rules[2]
- IBM style guide Vale implementation[3] Grafana
- GitLab's Vale implementation[4]
- Crossplane's Vale rules[5]
- openSUSE Vale style rules[6]
- MongoDB Vale rules and GitHub Action[7]
- MailChimp Vale rules[8]

While many of these organizations do similar checks, you'll find differences in specific implementations.

---

1. https://github.com/grafana/writers-toolkit/blob/main/vale/Grafana/styles/
2. https://redhat-documentation.github.io/vale-at-red-hat/docs/main/user-guide/redhat-style-for-vale/
3. https://github.com/errata-ai/IBM
4. https://gitlab.com/gitlab-org/gitlab/-/tree/master/doc/.vale?ref_type=heads
5. https://github.com/crossplane/docs/tree/master/utils/vale
6. https://github.com/openSUSE/suse-vale-styleguide
7. https://github.com/mongodb/mongodb-vale-action/tree/main
8. https://github.com/testthedocs/MailChimp

Thank you!

We hope you enjoyed this book and that you're already thinking about what you want to learn next. To help make that decision easier, we're offering you this gift.

Head on over to https://pragprog.com right now, and use the coupon code BUYANOTHER2025 to save 30% on your next ebook. Offer is void where prohibited or restricted. This offer does not apply to any edition of *The Pragmatic Programmer* ebook.

And if you'd like to share your own expertise with the world, why not propose a writing idea to us? After all, many of our best authors started off as our readers, just like you. With up to a 50% royalty, world-class editorial services, and a name you trust, there's nothing to lose. Visit https://pragprog.com/become-an-author/ today to learn more and to get started.

Thank you for your continued support. We hope to hear from you again soon!

The Pragmatic Bookshelf

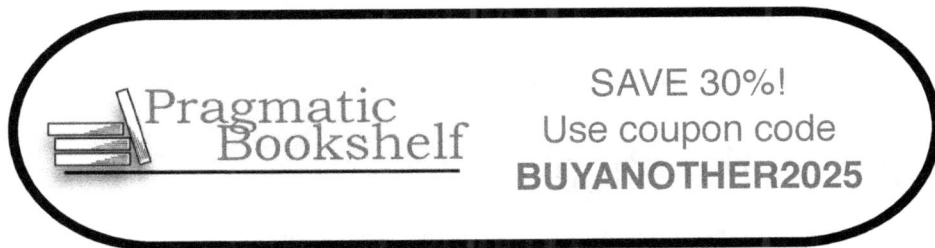

Pragmatic Bookshelf

SAVE 30%!
Use coupon code
BUYANOTHER2025

The Pragmatic Bookshelf

The Pragmatic Bookshelf features books written by professional developers for professional developers. The titles continue the well-known Pragmatic Programmer style and continue to garner awards and rave reviews. As development gets more and more difficult, the Pragmatic Programmers will be there with more titles and products to help you stay on top of your game.

Visit Us Online

This Book's Home Page
https://pragprog.com/book/bhvale
Source code from this book, errata, and other resources. Come give us feedback, too!

Keep Up-to-Date
https://pragprog.com
Join our announcement mailing list (low volume) or follow us on Twitter @pragprog for new titles, sales, coupons, hot tips, and more.

New and Noteworthy
https://pragprog.com/news
Check out the latest Pragmatic developments, new titles, and other offerings.

Save on the ebook

Save on the ebook versions of this title. Owning the paper version of this book entitles you to purchase the electronic versions at a terrific discount.

PDFs are great for carrying around on your laptop—they are hyperlinked, have color, and are fully searchable. Most titles are also available for the iPhone and iPod touch, Amazon Kindle, and other popular e-book readers.

Send a copy of your receipt to support@pragprog.com and we'll provide you with a discount coupon.

Contact Us

Online Orders:	*https://pragprog.com/catalog*
Customer Service:	*support@pragprog.com*
International Rights:	*translations@pragprog.com*
Academic Use:	*academic@pragprog.com*
Write for Us:	*http://write-for-us.pragprog.com*

www.ingramcontent.com/pod-product-compliance
Lightning Source LLC
Chambersburg PA
CBHW081746200326
41597CB00024B/4407